RELIGIOUS ENTHUSIASM AND THE GREAT AWAKENING

AMERICAN HISTORICAL SOURCES SERIES:
Research and Interpretation

LORMAN RATNER, Editor

PRENTICE-HALL INTERNATIONAL, INC., *London*
PRENTICE-HALL OF AUSTRALIA PTY. LTD., *Sydney*
PRENTICE-HALL OF CANADA LTD., *Toronto*
PRENTICE-HALL OF INDIA PRIVATE LTD., *New Delhi*
PRENTICE-HALL OF JAPAN, INC., *Tokyo*

David S. Lovejoy
University of Wisconsin

RELIGIOUS
ENTHUSIASM
AND THE GREAT
AWAKENING

Prentice-Hall, Inc., Englewood Cliffs, New Jersey

Current printing (last digit):
10 9 8 7 6 5 4 3 2 1

© 1969 by PRENTICE-HALL, INC.
Englewood Cliffs, New Jersey

Library of Congress Catalog Card No.: 70–86519

Printed in the United States of America
P: 13–773267–8, C: 13–773275–9

EDITOR'S FOREWORD

Religious Enthusiasm and the Great Awakening is a volume in the American Historical Sources Series, a series devoted to the exploration of aspects of American history and to the process of interpreting historical evidence. The introduction to each volume will be followed by some of the original documents used to prepare the essay. In this way readers are invited to share in the experience of turning raw evidence into history. The essay has been written especially for this series and is a contribution to historical knowledge as well as a demonstration of how history is written.

In writing of the Great Awakening, David S. Lovejoy lays bare the implications of a debate which not only had far-reaching effects in the 1740's, but which had continued significance for American religious history. The proponents of the Awakening, recognizing the increasing influence of Enlightenment thought and the serious decline in religious practice in the Colonies, sought to revive religion by urging Americans to take a more active personal and emotional part in the spread of God's grace. The revivalists warned of the consequences both to the individual and to society if religious complacency and disinterest continued to increase. They felt that grace came from God, but conversion was encouraged when people listened to those clergymen who claimed that they themselves had experienced God's grace and who, with His help, exhorted their audiences to recognize their sinful nature and open their hearts to Christ. Advocates of the Awakening regarded the era as a time of crisis in which a new approach to religious practice seemed necessary if the special promise of America, as a land whose people were in covenant with God and who served His purpose, were to be preserved.

But to many the Awakening seemed to pose its own threat both to religion and to the political and social stability of American society. At issue was whether such a revival was really the result of God's working outside the accustomed forms, infusing a holy spirit that would reinforce the role of religion and the Church in America, or whether it was enthusiasm, an utterly false impression heightened by an excess of human passion leading men to renounce traditional religious practices and abandon their churches. If we remember that the Church played a central role in American colonial life, we realize just how significant the Awakening was. In debating the means of religious practice in America, the participants touched upon the crucial question of the use of Old World institutions, as well as the continuance of Old World customs in the New World. It was a debate to be re-enacted at the time of the American Revolution and after, although the topic was expanded to include social and political institutions, practices, and traditions.

In describing the arguments as to whether the Great Awakening was enthusiasm and, as such, was to be decried, or an example of a new spirit and technique that was proper and necessary if the New World were to fulfill its promise, Professor Lovejoy has analyzed vital aspects of American history for us. He has provided the materials by which we can better understand these arguments and thus acquire a sense of the tone of the participants. We are now prepared to evaluate his analysis of the event and that is the purpose of this Series.

LORMAN RATNER

Herbert H. Lehman College
The City University of New York

CONTENTS

part two
A SPIRIT OF SUPERSTITION
AND ENTHUSIASM
61

Religious Enthusiasm
and the Great Awakening

ENTHUSIASM

The meaning of the word "enthusiasm" has not been constant through the years. Until recently, historically speaking, "enthusiasm" has had a religious connotation, which stemmed from the Greek word *enthousiasmos*—to be inspired or even possessed by a god or a divine, superhuman power. In the seventeenth and eighteenth centuries, the religious context of the word was retained, but its meaning was altered. Both Deists and Calvinists, for instance, sharply contrasted enthusiasm with revelation and inspiration. John Locke accepted the idea of revelation if it could be squared with reason or was reconcilable with Scripture, but if it were neither, it was patently enthusiasm, the product of man's "warmed or overweening brain." The third Earl of Shaftesbury defined inspiration as a "real feeling of the Divine Presence, and enthusiasm a false one." Enthusiasts, wrote Jonathan Edwards, are those "who falsely pretend to be inspired by the Holy Ghost as the prophets were."

By 1841 Emerson could write, "Everywhere the history of religion betrays a tendency to enthusiasm," a statement which tells us a good deal about the nineteenth century's break from both the rationalism of the Deists and the theological limits of Calvinism. At the same time, it tells us something about New England transcendentalists, whose romantic approach to religion depended upon inspiration—suspect a century before. Today, "enthusiasm" has lost much of its earlier meaning, for we use it commonly in positive fashion to describe any strong feeling of excitement or fervor in support of a subject or cause, religious or otherwise. But in the 1740's, the charge of "enthusiasm" was the ultimate reproach hurled against the revivalism of the Great Awakening.

The history of enthusiasm has been the history of "come-outers" * of one sort or another, whose religious thirst was not easily slaked, least of all by the forms of traditional religion. In their drive to know God, the come-outers were impatient with the confines of theology and ecclesiasticism, and they constantly sought a more direct, intense, and personal relationship with what their age and their contemporaries called God. Enthusiasts found rash action and vitality necessities for defying traditional religious restraint, and they insisted on emotional involvement in their search for righteousness. From the time of Christ to the present, enthusiasts have frequently set themselves apart, having discovered, they believed, the right view of things holy, cornering "truth," and defending it against religious convention.

In the seventeenth and eighteenth centuries there were many enthusiasts. The religious conflict of the English Civil War turned up not only Quakers—the most obvious enthusiasts—but also Ranters and Oliverians and a host of other splinter sects which responded sometimes violently, sometimes peacefully, to the commotion of their time. Louis XIV's persecution of the Huguenots eventually produced the French Prophets, or Camisards, whose frequent visions and trances kept religious enthusiasm alive in France for several years. Huguenots were welcome in England even before the revocation of the Edict of Nantes in 1685 when life across the Channel became dangerously uncomfortable. The French Prophets among them were tolerated as long as they kept out of trouble, but when, between revelations and transports, they attacked the Established Church as corrupt and unholy, English charity and sympathy wore thin. The best literary attack made on them was Shaftesbury's *Essay on Enthusiasm* (1711), although the Earl was doubtless interested in more important targets than a few shabby French visionaries, since his deistic writings opposed revealed

---

* Religious radicals who separate from organized churches, taking their cue from the Biblical charge in II Corinthians, 6: 17: "Therefore come out from them, and be separate from them, says the Lord. . . ."

religion in any form. Despite the large number of Huguenots who ventured to the colonies, the number of Prophets among them was small. One family, the Dutartes, appeared in South Carolina in the early 1720's, and they carried on in great style, stimulated by the preaching of a Moravian minister and the mystical writings of Jacob Boehme, a seventeenth-century German. Directed by impressions and impulses, visions and signs, the Dutartes' behavior finally became incorrigible, and they had to be restrained by the government. Their story ended sadly when two younger members were executed for murder.

Eighteenth-century Americans were well aware of religious enthusiasm by the time of the Great Awakening in the 1740's. No doubt only a few had read Shaftesbury's *Essay* or the strict warning on the subject in Locke's *Essay Concerning Human Understanding* (added to the 4th edition, published in 1700). Probably others had heard of Anne Hutchinson and the Antinomian Controversy of the 1630's, not many years after the Massachusetts Bay Colony was settled. Colonial society could easily brand Mistress Hutchinson an enthusiast, for she and her cohorts claimed an intimacy with God, a seal of the Spirit which rendered them sinless—a presumption no orthodox Christian would dare make. Mistress Hutchinson and her followers also believed in direct revelations from God. When she admitted as much at her trial, Governor John Winthrop's government banished her into the wilderness of Rhode Island where she preceded the Quakers by about twenty years.

It is not surprising, then, that advocates of the Great Awakening were called enthusiasts. Certainly the Revival brought with it radical religious expression, beliefs, and conduct not previously understood or appreciated. And once the Revival began to spread widely, a fundamental issue which split most American colonists into two camps was whether it was a genuine outpouring of the spirit of God or a first-rate example of religious enthusiasm.

We cannot ignore the hard truth that a religious revival occurred with a suddenness and an intensity never again equaled in America. Whether the Revival was genuine is a question which does not really concern us as historians today. What does concern us is how the people at the time felt about the Awakening and whether they thought it was genuine or false. Their response helps to explain why the Revival occurred and, at the same time, why it was surrounded by such intense emotionalism, an emotionalism which its champions defended as part of true religion and its opponents damned as enthusiasm, wild-eyed and destructive. The historical sources describing the Revival are very biased; such an issue, which divided American colonists as deeply as the Awakening, could leave few observers neutral. One was either for it or opposed to it; almost no one—except, perhaps, Benjamin Franklin— remained indifferent.

NEW WORLD, NEW BIRTH, NEW LIGHT

The causes of the Great Awakening have stumped historians for years. Most writers on the subject have had no trouble describing what happened, but why particular events occurred and what the consequences of these events were have been more difficult to determine. Some historians feel that the causes of the Great Awakening were international, for a wave of revivalism in one form or another swept across western Europe and Great Britain at about the same time or a little earlier. By the very vigor of this revivalism, it was bound to span the Atlantic and find an easy acceptance in the colonies. The best evidence we have of this theory is young George Whitefield, who first visited America in 1739, fresh from the upsurge of evangelical Methodism which John Wesley and he had strikingly preached in England and Scotland. Whitefield was doubtless a "cause" and a strong one, for he journeyed throughout the colonies, preaching and converting as he traveled.

Perhaps local and indigenous causes and conditions better explain the Awakening. For a variety of peculiar reasons, Whitefield encountered a people well prepared for an intense religious experience. Jonathan Edwards of Northampton, Massachusetts, had turned Solomon Stoddard's traditional "harvests" into a splendid revival in 1735, although it was local in scope. Far from burned out, these people and many others took fire again in 1740 at the time of the general awakening which was again led by Edwards in Massachusetts. Similarly, Theodorus Jacobus Freylinghausen, a Dutch Calvinist, and Gilbert Tennent, a Presbyterian, both warmly evangelical, were busy in the

1730's in New Jersey and Pennsylvania and had carefully prepared the way for Whitefield's preaching a few years later. Anglican Virginia was slow to respond to this movement, although the people of the backcountry were affected somewhat later by ardent revivalists sent by Pennsylvania Presbyterians of the Tennent stamp, the most renowned of whom was Samuel Davies. Brush-fire revivals which preceded the event no doubt give us a good deal of information, but do not explain completely the general awakening which blazed through British America during and after Whitefield's visit.

"A revival of Religion presupposes a declension." Doubtless this was as true in 1740 as it was a hundred years later when Charles Grandison Finney recorded the thought at the height of another revival. But whether it was actually true at the time is not as important as the fact that people believed it to be true. Edwards spoke often of the "general deadness all over the land" and of the "low state into which the Church of God has lately been sunk"; Whitefield complained frequently of the "degeneracy of the times." Gilbert Tennent attacked most of the clergy as "dead Formalists," and Samuel Davies in Virginia found not only "deadness" among the Anglican ministry, but corruption and ignorance as well. The Awakening was a reaction to religious decline—but why?

From the time of the very first settlements there was something special about America. Call this special quality the "Citty upon a Hill" idea, a "manifest destiny," or even the "Promise of American Life." Call it what you will, America felt it had a privileged place in God's plans, and colonists felt strongly about their peculiar Providence. Most noticeably, this idea was expressed as a covenant between God and His People, and although the Puritans of New England gave the most concrete form to this conception, the idea was to be found throughout the American colonies; in fact, it has never ceased to be an integral part of American history. God would be the people's Jehovah, and He would protect and guide them if the people lived according to His holy word and nurtured His grace among them. Where better to fulfill this covenant than in the New World, far from the ungodliness of the Old?

By 1740, thirteen colonies had carved territory out of the "howling wilderness"—several were already at least 100 years old—and the dangerous edge of the frontier had been made somewhat safer. The first half of the eighteenth century was a period of expansion and political, social, and economic development, about which all were satisfied. However, in the midst of material progress and prosperity, something seemed to be missing—something was not as it should be. For all its apparent success, America had fallen short of its promise.

Indeed, the Awakening was partly a reaction to the reason and rationalism which epitomized the Enlightenment. But it was not the Enlightenment itself which signaled a decline in religion as much as the complacency and smugness which had crept into eighteenth-century life along with it. In 1740 many people suspected that America was not worthy of its earlier promise, that it was degenerating like England and the Old World into materialism and self-satisfaction. Most notably, danger seemed to lurk in a rational approach to religion, tending to Arminianism and even Deism, which was offensive to most colonists whose theology and world view were essentially Calvinist. "Whatever is, is right" may have fitted Alexander Pope's sophisticated world in Augustan London, but it was not compatible with the religious attitudes of Puritans in New England and Scotch-Irish and Germans in the Middle Colonies and Virginia. For those who honored the covenant, or believed in a more vague, yet advantageous, conception of their relation with God, a slackening of religious endeavor was not only sinful, but disastrous, particularly because more was expected of men in America than of less privileged people elsewhere. That this was the "best of all possible worlds" may very well have satisfied enlightened rationalists in England and America, but it was far different from the conclusions reached by a good many colonists who were embarrassed by the religious decline they observed about them. The vital link with God which the New World had promised seemed to be missing. The seventeenth century had enjoyed it, or at least in retrospect had seemed to, but in the 1730's many people experienced a "general deadness," and God's spirit seemed to have been withdrawn. The Great Awakening was an attempt to do something about this state of affairs, to promote a return to the purity of an earlier condition when the grace of God abounded in the hearts of His people.

What many religious leaders actually did was to convince a large segment of the colonial population that not only was America different from the Old World, but religion itself was different in America. They held that a vital relationship demanded a "vital religion," and that a vital religion was an experimental religion which affected the hearts or emotions of the people. In fact, the emotions or "affections" were a vehicle for the new religion. This was a radical breakthrough for the colonial clergy who participated, for previously emotions had been carefully controlled, lest they be misdirected and lead to enthusiasm.

According to the new psychology, which Jonathan Edwards both understood and championed and which his colleagues found would work, colonial ministers of evangelical temperament appealed to the emotions of their people as the quickest way to conviction of sin and conversion. In this way they produced a revival which swelled into a general awakening. Another segment of the population, already succumbing to the rationalism of the age, soon accused the whole movement of enthusiasm and withdrew what support they had lent it. Of course, everyone damned enthusiasm, but the advocates of the Awakening refused to recognize the revivals they produced (or provoked) as anything but remarkable outpourings of God's grace. They insisted that this movement was something new and different which must be understood on its own terms, that extraordinary times demanded extraordinary methods. God in America was working outside the accustomed forms. After all, God could work in any way He chose, and in the New World He chose to work by a variety of means, some not acceptable to those who insisted on Old World, even seventeenth-century, criteria.

For the revivalists it was both a brave and a strange New World. For them, the movement was a revival of an original purpose and an awakening to new means of fulfilling this purpose. In spite of their theology and church order, which had previously restrained their zeal, they found a surprising quickening of religion through the "affections" or "passions" of their people. What to their opponents was enthusiasm and falsehood, was to the revivalists the unique characteristics of an extraordinary outbreak of divine grace. In the 1740's, American colonists were divided over the genuineness of the Awakening; the issue dividing them was enthusiasm.

THE NEW MEANS

If the proponents of the Awakening were radical in their emotional and dramatic approach to religion, they were, for the most part, orthodox in theology. After all, they were Calvinists at heart and had no doubt about the fundamentals of Christian doctrine. Man was damned; he needed grace for salvation; through Christ's death God provided grace for a few, and the rest went to hell. None of the revivalists felt that they had any cause to question these assumptions, although Whitefield tended to relax doctrinal rigidity. What the revivalists did question were the traditional *means* of grace, for these, they claimed, might be better utilized to promote God's plan for America. Thus, in handling these means, the revivalists were radical. The revival experience gave a new gloss to orthodox doctrine and practice. Since men were sinful, they must be convinced of it before they could expect God to reach out and save them. Therefore, revivalists stressed more strongly than ever the "terrors of the law" to bring men to an absolute surety that they were evil in God's eyes and deserved the death and punishment the Bible promised. Once convinced of this, despairing of any hope, they could look for no escape but to throw themselves on God's mercy—"shutup to Christ," as Tennent put it— wholly persuaded of their unworthiness to receive it. To bring about conviction, New Light* preachers painted stark pictures of hell and the likelihood of man's falling into it. They placed sharp emphasis on fire and brimstone and played heavily on the emotions of their listeners as the surest means to dissolve all support concealed in complacency and good works. Some ministers were more skillful than others in striking terror into the souls of the unregenerate. Gilbert Tennent was a "son of thunder" who had "learned experimentally to dissect the heart of a natural man." After hearing him preach, even Whitefield admired him and marveled and regretted his own softness in dealing with sinners. Edwards surpassed all other ministers in sheer ability with words; he bombarded the senses of his listeners with simple ideas of damnation and hell, well aware of the psychological effect upon their understanding. Conviction was a part of God's plan for conversion. The ministers were altogether willing to help Him promote it by literally scaring the hell out of their people. And scare them they did, producing at times mass physical reactions from whole congregations and a universal lament about what people could do to be saved. The intensity of the

---

* A term used generally to describe proponents of the Awakening. More specifically, it meant New England Congregationalists who supported the Awakening. Opponents were often called Old Lights.

reaction was often in proportion to the zeal of the minister. An extremist with a dramatic flair and a good pair of lungs might provoke outright frenzy among his listeners and come close to bringing down the roof.

Not all revivalists agreed about the efficacy of terror. Although Tennent used it indiscriminately, along with a canny ability to probe a sinner's heart, others such as Thomas Prince of Boston taught that conviction resulted "sometimes by words of terror, and sometimes by words of tenderness," that God could apply either to the conscience as needed so that the two methods "became as sharp arrows piercing into the heart."

To strike terror into the breasts of the unregenerate or to convict them by other means was not the only duty of the ministry, even in revival time, for ministers preached the Gospel as well. Without hope of reward, with hell staring him in the face, the sinner was prepared for God's act of grace, should He choose to bestow it, and the 1740's demonstrated that conversion was likely to come in connection with a revival and strong preaching. "Attend diligently upon the Word *Faithfully* dispensed," warned Tennent—which meant, of course, from revivalists like himself—for "Faith comes by hearing, and hearing by the Word of God." During the Awakening there was no dearth of the "Word," for churches and ministers kept land office hours. Even as some hearers were struck down, others were lifted up, and in meeting houses throughout the land the moanings and groanings of the condemned were offset by cries of ecstasy, even the holy laughter, of those experiencing "pangs of the New Birth."

Conversion in the 1740's, however, differed from what was usual and customary. It was frequently telescoped into a short period of time and was less a planting of a seed to be slowly nurtured by the believer than a "circumcision of the heart," as Edwards put it, or a cataclysmic encounter like Paul's on the road to Damascus. In what seemed to be a return to early Puritan practice, but with much sharper emphasis, the new saint regarded his conversion as an isolated event to be related and described as to time and place for the rest of his life. Under revival conditions the process of conversion became more hurried and concentrated. And inability to cite the precise nature of one's conversion was sometimes accepted as evidence of none at all.

The very nature of grace also tended to shift during the Revival. A seventeenth-century saint had an inkling which side of God he stood

on, but he never could be absolutely sure. Often after minor trans-
gressions he experienced a gnawing doubt as to the condition of his
soul and retired to his closet to search his heart for renewed evidence
of salvation. But those of the "New Birth" boasted confidence in grace.
The new character of conversion afforded the saint an assurance of salvation which at times smacked of the Antinomianism of Anne Hutchinson and was therefore heinous in the eyes of old-guard Calvinists. Once
he was sure of salvation, it was not much of a jump for a new saint,
having judged himself, to judge others, even ministers, on the state of
their souls, a practice which led to a good deal of resentment and often
schism and separatism.

Grace came from God, and deciding to whom He would grant it
was wholly His business. But New Light preachers were well aware
of their own role in the process of conversion, and on the basis of their
knowledge they established a new set of rules which released them
from old restraints and left them free to promote the Revival and the
grace of God. Previously, the state of a minister's soul had little effect
on his ability to preach the Gospel and perform his duties. Although
it was usually expected that a minister would be a saint, or at least experience conversion sometime during his ministry, if not before, the
chief criterion for a successful ministry was not the state of the minister's soul, but whether he preached sound doctrine. If he did, he contributed to the means of grace. But the Great Awakening changed all
this. "Experimental piety" demanded a minister who had already been
converted. To be a practitioner of vital religion, to aid in the process
of the New Birth, a minister must first have experienced grace, for
how could a dead soul promote what he was ignorant of? "We can
preach the Gospel of Christ no further than we have experienced the
power of it in our hearts," wrote Whitefield.

Since unconverted ministers were an obstacle to the warm flow
of grace, they were often censured. The New Light clergy, in promot-
ing God's work, took it upon themselves to distinguish between saints
and hypocrites among their colleagues. In such salubrious times the
clear light of grace enabled them to discern the state of another's soul
and, if he failed the test, to pronounce judgment on him. Cautious
Jonathan Edwards warned against this usurping of God's prerogative;
"To separate the sheep from the goats," he wrote, "is the proper business of the day of judgment." But there was no stopping extremists

such as James Davenport in New England and several of Tennent's followers in backcountry Virginia. They freely condemned those whom they judged as unconverted and, what is more, persuaded their congregations to forsake them as "carnal graceless wretches."

The loudest voice against an unconverted ministry was Gilbert Tennent's. At the height of the Awakening he preached and then published a bitter sermon aimed at "carnal" ministers who did not know God. Comparing these ministers to the Scribes and Pharisees who had opposed Christ, Tennent declared that "Pharisee Teachers will with utmost Hate oppose the very Work of GOD'S Spirit, upon the Souls of Men; and labor by all Means to blacken it." Whitefield found Tennent's argument "unanswerable"; a clergy without grace was the "bane of the Christian Church."

The "Nottingham Sermon" had more effect than Tennent bargained for. He and his "red-hot Bretheren" from the Log College had for several years been feuding with the Philadelphia Synod of the Presbyterian Church. The dispute grew more intense as the Revival strengthened and spread; most of the issues concerned the Synod's authority, but included in the dispute was the question of itinerant preaching and other New Side* activities. In his notorious sermon Tennent blasted unconverted ministers without mercy and very strongly implied that his opponents in the Synod, or anywhere, for that matter, were among them. The result was a troublesome schism in the Presbyterianism of the Middle Colonies. Forced to withdraw from the Philadelphia Synod, a few years later Tennent and his adherents established their own Synod, which included New York and New Castle.

The old idea of a minister's fixed relationship with his people and his parish was also discarded, at least among the revivalists. The terrors of hell and the grace of God had to be heightened by every possible means, and New Light clergymen did not let parish or colony boundaries confine their efforts. One of the most distinctive, and at the same time controversial, characteristics of the Great Awakening was the mobility of New Light ministers who, in their rush to deliver souls to God, became itinerants and preached, prayed, frightened, and consoled whenever and wherever an audience demanded. George Whitefield was the master itinerant; having first arrived in the colonies in

---

* In the Middle Colonies Presbyterian advocates of the Awakening were generally called New Side and their opponents, Old Side.

1739, in a few months' time he had visited them all, where his listeners by the thousands melted under the influence of the Word. The young Whitefield had come to Georgia, which had been settled only a few years earlier, to establish an orphanage for neglected children of the new colony's inept and indigent settlers. Seeking contributions from American colonists to support his good work, he journeyed north several times, preaching as he traveled, spreading his evangelical message to a people ripe for awakening. Whitefield set a pattern for itinerants. No sooner had he returned from New England when Gilbert Tennent traced his footsteps to Massachusetts "to *water the good Seed sown by Mr. Whitefield*," barely allowing Boston time to recover between visits. As the Awakening increased in intensity, its clergymen advocates moved from church to church and from town to town with or without invitations. As people flocked to hear them, their own parish ministers were often left without listeners, thus arousing suspicion about the state of their souls.

Had itinerant preaching attracted only the skills of ordained ministers, objections to it from the Revival's opponents might have been less severe. But the Awakening also encouraged exhorters whose credentials for their holy tasks were no more than heated devotion and a call to preach. Some thought that the prestige of the ministry suffered with rank amateurs loose in the field, but the criteria among the religiously aroused were not college degrees and proper ordinations, but grace and a zeal to promote it.

The same dispensation which dictated a new look at conviction and the "law," at conversion and the means of grace, that demanded a converted ministry and encouraged itinerating and exhorting, permitted a change in traditional ideas of education and ministerial control, particularly if they tended to inhibit the Revival. Although previously a minister needed a degree from a European university or a recognized colonial college, New Lights accepted products of the Log College at Neshaminy, Pennsylvania, where, in reaction to the formalism and head-learning of the universities, Gilbert Tennent's father had trained a generation of revival preachers for what he felt really mattered and sent them out to awaken the world. Opponents of the Awakening, in addition to insisting upon formal learning, demanded a proper ordination according to traditional ecclesiastical procedures. No wonder they looked askance upon Isaac Backus of Norwich, Connecticut, who, a year or two after his conversion, felt called upon by God to preach the Gospel and promptly did, itinerating, exhorting, and finally settling at Middleborough, Massachusetts, after being called there to shepherd a separatist congregation.* Backus was minister at Middleborough for over sixty years, the only change in arrangements being that he led a shift to the Baptist Church with most of his congregation a few years after his arrival. In the opinion of Old Lights and Anglicans he was an ignorant, exhorting nobody without education or a rightful claim to pursue God's work. That he championed the separation of church and state, a conviction he formed from his experience as a separatist, did not endear him to Congregational Massachusetts, where public worship of God had been protected by law since the 1630's.

It would be difficult to equate New Lightism with anti-intellectualism except in a few special cases such as Davenport's book burning at New London. Backus' splendid *History of New England* in three volumes certainly belied the charge that he opposed intellectual activity. Most New Lights accepted learning as part of a minister's "baggage," as long as he never emphasized it at the expense of the heart. "I love to study," wrote Whitefield, "and delight to meditate . . . and yet would go into the Pulpit by no Means depending on my Study and Meditation, but the blessed Spirit of God." The majority of revivalists undoubtedly preferred a nice balance between intellect and emotion, but as itinerants and exhorters quickened the tempo, their emphasis became less that of learning and more the heart and its needs. Tennent, his critics complained, "roar'd more fiercely" than Whitefield "against Colleges, Human Reason and Good Works."

If, as Whitefield suggested, he entered the pulpit full of the "Spirit of God," there was no need, it seemed, for written sermons. "Extempore preaching" became as much a part of the revivalists' style as an appeal to the "affections"; in fact, they were intimately related. A minister read the pulse, or sensed the mood, of his listeners and tailored his message accordingly. The "Spirit of God" moved quickly and a clergyman had to stand ready and move with this spirit, not be tied to a stuffy sermon written in the cool of his study. A sermon had to smell of brimstone, not of the lamp. In the heat of the Awakening a good many ministers forgot their written sermons, even their notes, and preached on the spot

---

* Separatists were supporters of the Revival who withdrew from their accustomed churches and formed their own congregations.

as the spirit moved them in an extemporaneous manner. Their method was both exciting and effective, lending to their sermons and prayers an emotional appeal which was the very essence of revival.

"ENTHUSIASTICAL SCHISMATICKS"

Proponents of the Awakening insisted that revival activities were a manifestation of Christ's Spirit upon the land, exempting them from customary ecclesiastical restraints, sometimes even scriptural guidelines. Opponents, whether Old Light, Old Side, rationalist, or Anglican, stigmatized these same activities as fanaticism and enthusiasm, and lumped together as enthusiasts moderates such as Edwards, the extremist Davenport, and contentious Gilbert Tennent.

Old Lights and the Old Side objected to the Awakening on other grounds. Certainly there was honest disagreement over the nature of religion, the meaning of conversion and grace, and the efficacy of reason and learning. There were decorum and tradition to be considered. Moreover, there were the established means through which God's grace worked, and anything else was considered suspect, particularly if it opposed the principles of logic and formal theology. Perhaps more important was the Revival's affront to the established clergy, who were often looked upon with contempt and whose offices and duties were frequently disregarded when itinerants swooped down on likely spots and congregations flocked to hear them.

There is no doubt that the Awakening disrupted organized churches, whether Congregational in Massachusetts and Connecticut, Presbyterian in Pennsylvania and New Jersey, or Anglican in Virginia. Whitefield generally ignored denominational lines; although he was nominally an Anglican, he preached mainly in dissenting churches to mixed audiences or in the fields and market places where ecclesiastical divisions meant little. Moreover, he frequently spoke in ecumenical terms, lamenting the differences between dissenters and conformists: "Oh, that the partition-wall were broken down, and we all with one heart and one mind could glorify our common Lord and Saviour Jesus Christ!"

The established denominations fought back as best they could against the blending of congregations and revivalists' practices which disturbed the accustomed order. Most severe was the response in Connecticut, where the Old Light legislature restricted the activity of foreign and domestic itinerants. Samuel Davies in Virginia had to contend not only with the colony government but also with the Bishop of London and his Commissary in order to keep revival channels open in Hanover County and other backcountry places. Had Davies preached only to Presbyterians, the government might have left him alone, but many of his listeners were former Anglicans—with more coming over every day —who were bored with the "languid harangues" and "insipid speculations" of the Church of England. The Massachusetts government took action against men such as Davenport whose conduct, according to most Yankees, constituted a public nuisance. Accused of "maliciously publishing . . . slanderous and reviling Speeches," Davenport was tried by the Superior Court; rather than find him guilty, it declared him *Non Compos Mentis* and gladly sent him away.

A widespread reason for opposing the Revival was the belief that it posed a threat to social and political stability. Tracts, sermons, letters, and newspapers condemned the "New Way" and spoke of the ignorant, the rabble, the "admiring Vulgar," and Negroes, all of whom the preachers and exhorters aroused and made restless and kept from their callings. Virginia Anglicans accused Davies of *"holding forth* on working days," contrary to the "religion of labor," and causing good Virginians to neglect their duty in providing for their families. Davies certainly did not increase his popularity in either Williamsburg or London when he replied that his people spent less than half as many working days listening to him hold forth on the "Word of Life" as Anglicans were "obliged to keep holy according to their calendar." Connecticut's government claimed that Davenport's "conduct and doctrine" had a "natural tendency to disturb and destroy the peace and order" of the colony. That the "peace and order" of the American colonies generally were disrupted by the Awakening, there can be no doubt. How seriously upset and what the consequences are questions which historians today are only beginning to answer.

Those hostile to the Awakening had various reasons for opposing it, ranging from contrary views of religious temperament to fears of encroachment upon privilege and class. However, all the opponents of the Awakening agreed upon its enthusiasm. No charge against the Revival and its participants was more frequent; to no other charge was the "New Way" more vulnerable. According to the critics, God worked in accustomed and reasonable ways as the Bible taught and ministers had preached for some time. The use of terror, they claimed, played upon people's fears rather than convicted them of sin; it thus tended to drive sinners from Christ rather than toward Him. What passed among revivalists for conversion experience, no matter how vivid and dramatic, was considered contrary to God's usual means and opened the door to impulses and visions which impressed the gullible but cheapened the revealed word in the Bible. No man was more presumptuous than he who was overconfident of obtaining grace and judged his neighbors and the clergy on his own terms, not God's, at the same time claiming holy guidance. The ministry was a noble calling and instrumental in actually saving souls. Therefore, when New Light clergymen established themselves as a "holy band," separating themselves and their people from those whom they rashly judged and called unconverted, this was thought to be the height of pride and arrogance which could only lead to factionalism, schisms, and contempt. Itinerants and exhorters distracted the people, upset the traditional churches, and were "the destroyers of good learning and gospel order." And all this was done from a "false spirit," said Jonathan Mayhew, by "enlightened Ideots" who made "inspiration, and the Spirit of truth and wisdom, the vehicle of nonsense and contradictions."

Critics with a historical bent delighted in describing the Awakening against a background of enthusiasm culled from "all Parts and Ages of the world." The Anglican Commissary of Charleston, South Carolina, in his ridicule of Whitefield, ransacked "church history for instances of enthusiasm and abused grace." And with the "grand itinerant" in a pew before him, he compared him to "all the Oliverians, Ranters, Quakers, French Prophets," and even the celebrated Dutartes family who had outraged Carolinians a few years earlier with their trances and visions, their incests and murders. A Salem Anglican wrote home to England that the illusions of the French Prophets were nothing to the "confusion, disorder, & irregularity" of Boston, and if something very soon did not put an end to it, the year would be as memorable for enthusiasm as 1692 was for witchcraft. The converted, he wrote, "cry out upon the unregenerated, as the afflicted did then upon the poor innocent wretches [that] unjustly suffered."

The French Prophets were a favorite comparison, probably because their notorious activities were still within the memory of a good many people in the 1740's. Boston's Charles Chauncy was doubtless the most learned—and the most bitter—critic of the Revival, and he taunted New Lights with the Prophets again and again. In 1742 there was published in Boston a lengthy tract called *The Wonderful Narrative* which faithfully traced the history of the Huguenot extremists; in a revealing Appendix the author—who may or may not have been Chauncy— related what he saw at his doorstep in Boston to the *"Raptures and Ravishments"* of the French fanatics. Chauncy's most significant work, a volume called *Seasonable Thoughts on the State of Religion in New England*, which was published the following year, was a full-fledged attack upon Edwards and the Awakening and the extraordinary use the New Lights made of the various means of grace. "The Plain Truth," Chauncy declared, "is [that] an enlightened Mind, not raised Affections, ought always to be the Guide of those who call themselves Men." The effect of divine truth upon the soul should be a reasonable "Solicitude," not the wild shenanigans of New Lights. Chauncy was the Old Lights' rational man, and his book was a careful exposure of the Great Awakening to the light of reason. Moreover, in addition to its criticism, Chauncy's volume was a splendid explanation of what the revivalists were rebelling against.

"LUKEWARMNESS IN RELIGION IS ABOMINABLE"

It was much easier to make fun of the Awakening and hurl charges of enthusiasm than defend it. A positive view of the Revival was difficult to impress upon its critics since most were religious rationalists like Chauncy who delighted in ridiculing emotion, or Anglican moralists like Whitefield's Commissary, for whom religion was formal and professional. At the same time, the hard-core New Lights, who found an enduring spiritual essence in the Awakening, were forced to protect this essence from fanatics within their own group like Davenport who made fools of themselves. The very attempt to guard the Awakening from these people was accepted as an admission that enthusiasm *did* accompany the Revival. The supporters of the Awakening had the difficult job of arguing that, despite the lunatic fringe, the heart of the Awakening was pure and beyond all argument "a most Wonderful Work of God's Grace."

The defense went about its tasks in several ways. One method was to deny outright that the movement was centered on enthusiasm and to brand all who claimed it was as carnal and unconverted and enemies to God. At the height of the Awakening this was Tennent's chief weapon, and in his hands it did little to convince his opponents that they were mistaken. Another means of defending the Awakening was to cite Biblical precedents for alleged irregularities. The lamentations, the groanings, the faintings, said Tennent, were "no more than what the Scriptures inform us did happen in the apostolick times." Whitefield was frequently called an enthusiast. Harvard College accused him of speaking like a man who believed he had direct communications from God and as much intimacy with Him "as any of the Prophets and Apostles." Whitefield answered the Harvard people guilelessly: Certainly he had communion with God, he said, "to a degree," and had he not, he never would have become a minister. In what other way are saints chosen, he asked? The charge of enthusiasm missed its mark on Whitefield, for he claimed no "false spirit" but declared his communion with God to be the real thing.

Jonathan Edwards has left us the most profound defense of the Awakening. One of his efforts was a commencement sermon which he delivered at Yale College in the spring of 1741 called "The Distinguishing Marks of a Work of the Spirit of God." In this sermon Edwards argued calmly and subtly that, although physical manifestations were not necessarily a sign of the Spirit of God, neither were they necessarily *not* a sign. What Edwards attempted to do, and what most advocates of the Awakening had difficulty accomplishing, was to place the Awakening in an acceptable perspective as an extraordinary work of God. It was clear from Scripture, Edwards told his Yale audience, that God had certain things He wished to accomplish which had never yet been seen. Therefore, deviations from what men understood to be usual ought not to be used as evidence of enthusiasm, that is, if they did not violate His prescribed rules. "We ought not to limit God," Edwards wrote, "where he has not limited himself." The extraordinary means used by God at the time could very easily produce "extraordinary appearances." Given the weakness of man's nature, these "effects" should be expected and excused, not made the center of controversy and condemned wholesale by the doubters. If we look back upon church history, wrote Edwards, "we shall find no instance of any great revival of

religion, but what has been attended with many such things." He held
that physical effects had little to do with what was really going on and
therefore should not be used to support or attack the Awakening. Gilbert Tennent grasped this part of the argument and although he undoubtedly was one of the worst offenders against a reasonable "Solicitude" and decorum, he and his Log College friends could ask,

> What if there were some things exceptionable in the conduct
> of some of the instruments and subjects of this work? Is this so
> strange an incident in a state of imperfection, as to give us ground
> of surprize or prejudice against the whole work?

Proponents of the Awakening found themselves on the defensive
for several years. Surprisingly, Tennent and Davenport shifted courses
to preserve the purity of the movement. Friends had convinced them
that the Awakening suffered seriously from extremism. In order to res-
cue the substance of the Revival, they argued, the crackpots should be
quickly isolated. By 1744 both Tennent and Davenport had confessed
their "misguided zeal" and retreated from a number of the militant po-
sitions taken earlier and defiantly held. Tennent spoke first and included a severe condemnation of Davenport. But it was too late. Opponents of the Awakening seized upon the confessions as admissions not
only of enthusiasm but also of guilt, which more fully convinced them
of the disingenuousness of the movement.

The charge of enthusiasm dogged Edwards until his death in 1758.
At times he seemed as interested in protecting "true experimental religion" from some of the revivalists as in defending both from the charge
of enthusiasm. For a number of reasons, in 1747, Edwards chose to review the life of his friend David Brainerd, a young New Light minister of Connecticut who had died that year in his thirties. Since Brainerd
was widely known as a saintly minister and had been an active revivalist during the Awakening—although he had been converted a year
or two earlier—Edwards found his career to be a perfect example of the
good "effects" of experimental religion. He used Brainerd's life—almost
exploited it—to distinguish the "New Birth" from enthusiasm by edit-
ing Brainerd's memoirs, adding a revealing commentary, and publishing them together in 1749.

Edwards was certain that the life of his friend proved there was indeed a truly vital religion, a religion which arose from "immediate divine influences, supernaturally enlightening and convincing the mind, and powerfully impressing, quickening, sanctifying and governing the heart." The fact that "immediate divine influences" were evident in Brainerd's life was reason enough to accept these influences even though they seemed contrary to traditional means God used in dealing with His saints. And this was true despite many "pretences" and "appearances" in others which had "proved to be nothing but vain, pernicious enthusiasm." After reading his memoirs, if anyone still insisted that Brainerd's religion was enthusiasm, a "strange heat," a "blind fervor," Edwards then asked: Were his *virtues* the "fruits of enthusiasm?" (Here he listed about a page of these virtues, really an Edwardsean catalogue of the attributes of grace.) If these were the fruits of enthusiasm, "why should not enthusiasm be thought a desirable and excellent thing? For what can true religion, what can the best philosophy do more?" If "vapors and whimsey" make men thoroughly virtuous and lead them to the "most benign and fruitful morality," helping men to maintain both through a life of many trials while in the meantime opposing the "wildness, the extravagances, the bitter zeal, assuming behavior, and separating spirit of enthusiasts" (all the bad "effects" of the Awakening), if vapors and whimsey can do this, Edwards declared, the world has cause to "prize and pray for this blessed whimsicalness, and these benign sort of vapors!"

The logic here is tricky, and Edwards's readers might easily miss the point. Far from condoning enthusiasm, Edwards shared his century's mistrust of it. Nor was he flippantly asking, "If this be enthusiasm, who needs grace?" for he found no time to be amusing. The key to the argument is probably the phrase, "the fruits of enthusiasm," which Edwards used twice in this passage, along with a "fruitful morality" maintained "through a course of life." If the fruits of enthusiasm are "true virtue," then there is working in the converted person something quite apart from enthusiasm as it was understood at the time— obviously, the grace of God, which supported these virtues throughout life in the saint's pursuit of His kingdom. Others may see "vapors and whimsey," but if the fruits are the virtues of a Brainerd, let the world call "true experimental religion" what it will, even enthusiasm.

It was left to Edwards to salvage the meaning of the Awakening for New Lights who survived him and the evangelical generations which followed. Despite the century's penchant for rationalism, the "affections," said Edwards, primarily those related to "fear and sorrow, desire, love, or joy," were *bona fide* channels for transmission of religious truth. That an appeal to these emotions produced what some people called enthusiasm and occasionally "extraordinary effects on persons' bodies," meant little in Edwards's large view of God's plans. What mattered was that the new "means" encouraged conviction and conversion and therefore grace and virtue, and it encouraged them sweepingly throughout the colonies in the form of an Awakening. Related to Edwards's understanding of the remarkable outpouring of grace was his conception of America's peculiar Providence in God's over-all scheme for redemption and "enlargement of his kingdom." His lucid explanations of conviction and conversion, of grace and true virtue, were laced with a millennialism, a manifest destiny which, he claimed, the Awakening foreshadowed. God would work with extraordinary means in America, and the Awakening was His sign. Who could doubt the auspicious beginning? Only fools and the ignorant, or perhaps the rationalists, would be put off by "extraordinary effects."

Although American colonists could agree generally that enthusiasm was a "false spirit" and therefore evil, they could not agree whether the Revival represented a true spirit or enthusiasm. However, whether the colonists approved or disapproved of the Awakening depended upon their own religious temperament, on the manner in which they believed God manifested Himself in human society, and on what they understood to be the meaning of their experience in the New World. Those who accepted America intellectually and religiously as a logical extension of the Old World and applied Old World criteria to questions of faith, who believed that theology and religious institutions, as they were, gave reasonable form to the means of grace and church government, these people were shocked and offended by the monstrous irregularities of the Awakening. They pronounced it damnable enthusiasm.

For others, America was a radical chapter in the history of God's Providence. They demanded a vital link between themselves and the God who sustained them in the new experience. Traditional religious convention frustrated their emotional drive, a romantic urge, to press close to God, to experience the "New Birth" which the New World promised. To these people new means were necessary. To them the

Awakening was a blessing, it was far from false, it was not enthusiasm.

Neither side had the last word. Revivalism has endured in America as has the rationalism of Chauncy and the moralism of Anglicanism. Although today the zeal of revivalism often lacks the sharpness Whitefield and Edwards and Tennent gave it, one can still say with Emerson: "Everywhere the history of religion betrays a tendency to enthusiasm."

part one

A MOST
WONDERFUL WORK
OF GOD'S GRACE

The Grand Itinerant

George Whitefield

> George Whitefield's "deep-toned" and "melodious voice," his master-
> ful oratory, and sense of drama, even melodrama, made him one of
> the most effective revivalists of all time. He was in his mid-
> twenties when he first preached in the colonies, and he carried
> with him nearly all his listeners. However, there were some who
> opposed him, who were offended by his presumptuous intimacy
> with God. Whitefield kept close account of the Awakening, a kind
> of running description which he published serially as *Journals*. An
> indefatigable preacher, he was undoubtedly an immediate cause
> of the Awakening.

PHILADELPHIA, BURLINGTON, AND TRENT TOWN
IN THE JERSEYS

Monday, November 12. . . . A man came to me this morning,
telling me what God had done for his soul, by my preaching of faith.
He seemed deeply convinced of sin, and said he was drawn by God's
Spirit to pray last night, for which he was immediately looked upon
by his master and the family as a madman. I never yet knew one truly
awakened, who did not commence a fool for Christ's sake. I have great
reason to believe a good work is begun in many hearts. Lord, carry it
on for Thy dear Son's sake. At my first arrival at Philadelphia, I received a letter, which had been left for me three months, and in which
was a pressing invitation, sent me by Mr. Noble, (a spiritual man), on
behalf of many others, to come to New York. On Friday, I received another from the same person; which, looking like the call given St. Paul,
when the man appeared to him, saying, "Come over to Macedonia, and
help us," I, this morning, set out for that place. Four horses were lent
to me and my friends; and more we might have had, had there been

George Whitefield's Journals (London: The Banner of Truth Trust, 1960),
pp. 346–48, 441–45, 469–71, 473–74, 481–83.

occasion. About one, we got safe to Burlington, in the Jerseys, (twenty miles from Philadelphia), where I was importuned to preach as I went along. Immediately after dinner, I read prayers and preached in the church, to a mixed but thronged and attentive congregation. The poor people were very importunate for my staying with them all night, and giving them another discourse; but it being inconvenient with my business, with great regret, about five in the evening, we took our leave, and by eight o'clock reached Trent, another town in the Jerseys. It being dark, we went out of our way a little, in the woods; but God sent a guide to direct us aright.

Tuesday, November 13. Left Trent at six in the morning, and reached Brunswick, thirty miles distant, at one. Here we were much refreshed with the company of Mr. Gilbert Tennent, an eminent Dissenting minister, about forty years of age, son of that good old man who came to see me on Saturday at Philadelphia. God, I find, has been pleased greatly to own his labours. He and his associates are now the burning and shining lights of this part of America. Several pious souls came to see me at his house, with whom I took sweet counsel. At their request, and finding there was a general expectation of hearing me, I read the Church Liturgy, and preached in the evening at Mr. Tennent's meeting-house; for there is no place set apart for the worship of the Church of England, and it is common, I was told, in America, for the Dissenters and Conformists to worship at different times, in the same place. Oh, that the partition-wall were broken down, and we all with one heart and one mind could glorify our common Lord and Saviour Jesus Christ!

At my first getting up, I was somewhat weak, but God renewed my strength, and enabled me to speak with freedom and power. I was above an hour in my sermon, and I trust I shall hear it was not preached in vain. Paul may plant, Apollos water; Thou, Lord, only canst give the increase.

Wednesday, November 14. Set out from Brunswick, in company with my dear fellow-travellers, and my worthy brother and fellow-labourer, Mr. Tennent. As we passed along, we spent our time most agreeably in telling one another what God had done for our souls. He recounted to me many instances of God's striving with his heart, and how grace, at last, overcame all his fightings against God. About noon, we got to Elizabeth Town, twenty-two miles from Brunswick.

Here we took boat, and at four reached New York, where we were most affectionately received by the family of Mr. Noble. I waited upon Mr. Vessey, the Commissary, but he was not at home. Then I went to the meeting-house to hear Mr. Gilbert Tennent preach, and never before heard such a searching sermon. He convinced me more and more that we can preach the Gospel of Christ no further than we have experienced the power of it in our own hearts. Being deeply convicted of sin, by God's Holy Spirit, at his first conversion, he has learned experimentally to dissect the heart of a natural man. Hypocrites must either soon be converted or enraged at his preaching. He is a son of thunder, and does not fear the faces of men. After sermon, we spent the evening together at Mr. Noble's house. My soul was humbled and melted down with a sense of God's mercies, and I found more and more what a babe and novice I was in the things of God. Blessed Jesus, grant I may make continual advances, until I come to a perfect man in Thee.

. . .

CHRIST'S CHURCH, CHARLESTON, AND JOHN'S ISLAND

Saturday, July 12. Went over the water on Thursday, and read prayers and preached at the request of the churchwardens and vestry, at Christ's Church. Returned in the evening to Charleston; preached twice there yesterday, and went this morning to John's Island, about twenty miles up the river, whither I was invited by Colonel G—s. He received me and my friends most hospitably, and provided several horses, chairs, etc., for us and his family. We rode to the church, where there was a great congregation. God strengthened me to read prayers and preach twice with much freedom. About four we returned to the Colonel's. I was enabled to give a warm and close exhortation to the rich that sat about me, and returned to town in the evening, praising and blessing God. Glory be to His Most Holy Name, Dagon seems daily to fall before the ark. A lasting impression, I am persuaded, is made on many hearts; and, God, I believe, will yet shew that He hath much people in Charleston, and the countries round about. Lord, Thou hast visited them with Thy judgments, melt them down with Thy mercies. Stretch out the golden sceptre of Thy favour, and bruise them no more with Thy iron rod. Amen.

Sunday, July 13. Preached this morning, and collected in the evening for my poor orphans. Great numbers stood without the doors, and, it raining very hard during the time of Divine service, many of them were driven away. However, God caused the other people's hearts to devise liberal things. Upwards of £50 sterling was collected on the occasion, most shewing a readiness to assist me.

In the morning I went to church, and heard the Commissary preach. Had some infernal spirit been sent to draw my picture, I think it scarcely possible that he could have painted me in more horrid colours. I think, if ever, then was the time that all manner of evil was spoken against me falsely for Christ's sake. The Commissary seemed to ransack church history for instances of enthusiasm and abused grace. He drew a parallel between me and all the *Oliverians, Ranters, Quakers, French Prophets,* till he came down to a family of the *Dutarts,* who lived, not many years ago, in South Carolina, and were guilty of the most notorious incests and murders. To the honour of God's free grace be it spoken, whilst the Commissary was representing me thus, I felt the Blessed Spirit strengthening and refreshing my soul. God, at the same time, gave me to see what I was by nature, and how I had deserved His eternal wrath; and, therefore, I did not feel the least resentment against the preacher. No; I pitied, I prayed for him; and wished, from my soul, that the Lord would convert him, as He once did the persecutor Saul, and let him know that it is Jesus Whom he persecutes. In the evening, many came, I was informed, to hear what I would say; but as the Commissary hinted that his sermons would be printed, and as they were full of invidious falsehood, I held my tongue, and made little or no reply. . . .

Sunday, July 20. Preached in the morning as usual, and went afterwards to church, to hear the Commissary. His text was, "Take heed how ye hear." At first, I thought we should have a peaceable sermon, especially, since we had conversed the night before so amicably; but the gall soon began to flow from his tongue, though not with such bitterness as last Sunday. He endeavoured to apologise for his proceedings, condemned all that followed me, and gave *all* hopes of Heaven who adhered to him and the Church. In the evening (though I went off my bed to do it, and was carried in a chaise) the Lord Jesus strengthened me to take my last farewell of the people of Charleston. Many seemed to sympathise with me.

Blessed be God for sending me once more among them. Though the heat of the weather, and frequency of preaching, have perhaps given an irrecoverable stroke to the health of my body; yet, I rejoice, knowing it has been for the conviction, and I believe conversion of many souls. Glory be to God on high, the fields here, as well as elsewhere, are now white, ready to harvest. Numbers are seeking after Jesus. Two or three Dissenting ministers, by my advice, agreed to set up a weekly lecture. I advised the people, since the Gospel was not preached in the church, to go and hear it in the meeting-houses. May the Lord be with both ministers and people, and cause them to preach and hear as becomes the Gospel of Christ. At my first coming, the people of Charleston seemed to be wholly devoted to pleasure. One, well acquainted with their manners and circumstances, told me that they spent more on their polite entertainments than the amount raised by their rates for the poor. But now the jewellers and dancing-masters begin to cry out that their craft is in danger. A vast alteration is discernible in ladies' dresses; and some, while I have been speaking, have been so convinced of the sin of wearing jewels, that I have seen them, with blushes, put their hands to their ears, and cover them with their fans. But the reformation has gone further than externals. Many moral, good sort of men, who before were settled on their lees, have been awakened to seek after Jesus Christ; and many a Lydia's heart hath the Lord opened to receive the things that were spoken. Indeed, the Word often came like a hammer and a fire. Several of the negroes did their work in less time than usual, that they might come to hear me; and many of the owners, who have been awakened, have resolved to teach them Christianity. Had I time and proper schoolmasters, I might immediately erect a negro school in South Carolina, as well as in Pennsylvania. Many would willingly contribute both money and land. Almost every day something was sent for my orphans at Georgia. The people were very solicitous about my health, when they saw me so weak, and sent me many small presents. I sometimes feared they would be too hot against the Commissary; but I endeavoured to stop their resentment, as much as possible, and recommended peace and moderation to them, in most of my discourses. May the Lord Jesus reward them for all their works of faith, and labours which have proceeded of love. May He never leave them without a teaching priest, and grant that the seed sown in their hearts may grow up into an eternal harvest! Amen and Amen!

.   .   .

BOSTON

Thursday, October 9. Every morning, since my return, I have been applied to by many souls under deep distress, and was grieved that I could not have more time with them. Gave, this morning, the public lecture at Dr. Sewall's meeting-house, which was very much crowded. When I came near the meeting-house, I found it much impressed upon my heart, that I should preach upon our Lord's conference with Nicodemus. When I got into the pulpit, I saw a great number of ministers sitting around and before me. Coming to these words, "art thou a master in Israel, and knowest not these things?" the Lord enabled me to open my mouth boldly against unconverted ministers; for, I am persuaded, the generality of preachers talk of an unknown and unfelt Christ. The reason why congregations have been so dead is, because they had dead men preaching to them. O that the Lord may quicken and revive them! How can dead men beget living children? It is true, indeed, that God may convert people by the devil, if He chooses; and so He may by unconverted ministers; but I believe, He seldom makes use of either of them for this purpose. No: He chooses vessels made meet by the operations of His Blessed Spirit. For my own part, I would not lay hands on an unconverted man for ten thousand worlds. Unspeakable freedom God gave me while treating on this head.

After sermon, I dined with the Governor, who seemed more kindly affected than ever. He told one of the ministers, who has lately begun to preach *extempore*, "that he was glad he had found out a way to save his eyes." Oh that others would follow him. I believe, they would find God ready to help and assist them. In the afternoon, I preached, on the common, to about fifteen thousand people, and collected upwards of £200 for the orphans. Just as I had finished my sermon, a ticket was put up to me, whereon I was desired to pray "for a person just entered upon the ministry, but under apprehensions that he was not converted." God enabled me to pray for him with my whole heart. I hope that ticket will teach many others not to run before they can give an account of their conversion; if they do, they offer God strange fire. Went to a funeral of one belonging to the Council; but do not like the custom at Boston of not speaking at the grave. When can ministers' prayers and exhortations be more suitable, than when the corpse before them silently assists them, as it were; and, with a kind of dumb oratory, bids the spectators consider their latter end? When the funeral was over, I went to the almshouse, and preached on these words, "The poor received the Gospel," for near half an hour; then I went to the work-house, where I prayed with and exhorted a great number of people, who crowded after me, besides those belonging to the house, for near an hour more; and then, hearing there was a considerable number waiting for a word of exhortation at my lodgings, God strengthened me to give them a spiritual morsel. Soon after, I retired to rest. Oh, how comfortable is sleep after working for Jesus! Lord, strengthen me yet a little longer, and then let me sleep in Thee, never to awake in this vain world again.

• • •

Boston is a large, populous place, and very wealthy. It has the form of religion kept up, but has lost much of its power. I have not heard of any remarkable stir for many years. Ministers and people are obliged to confess, that the love of many is waxed cold. Both seem to be too much conformed to the world. There is much of the pride of life to be seen in their assemblies. Jewels, patches, and gay apparel are commonly worn by the female sex. The little infants who were brought to baptism, were wrapped up in such fine things, and so much pains taken to dress them, that one would think they were brought thither to be initiated into, rather than to renounce, the pomps and vanities of this wicked world. There are nine meeting-houses of the Congregational persuasion, one Baptist, one French, and one belonging to the Scots-Irish. There are two monthly, and one weekly lectures; and those, too, but poorly attended. I mentioned it in my sermons, and I trust God will stir up the people to tread more frequently the courts of His house.

One thing Boston is very remarkable for, viz., the external observance of the Sabbath. Men in civil offices have a regard for religion. The Governor encourages them; and the ministers and magistrates seem to be more united than in any other place where I have been. Both were exceedingly civil during my stay. I never saw so little scoffing, and never had so little opposition. Still, I fear, many rest in a head-knowledge, are close Pharisees, and have only a name to live. It must needs be so, when the power of godliness is dwindled away, where the form only of religion is become fashionable amongst people. However, there are "a few names left in Sardis, which have not defiled their garments." Many letters came to me from pious people, in which they complained of the degeneracy of the times, and hoped that God was about to revive His work in their midst. Even so, Lord Jesus, Amen and Amen. Yet Boston people are dear to my soul. They were greatly affected by the Word, followed night and day, and were very liberal to my dear orphans. I promised, God willing, to visit them again when it shall please Him to bring me again from my native country. The Lord be with thy ministers and people, and grant that the remnant, which is still left according to the election of grace, may take root and bear fruit, and fill the land!

.   .   .

STANFORD AND RYE

Wednesday, October 29. Came hither last night in safety, though dark and rainy. Was somewhat dejected before I went from my lodgings, and distressed for a text after I got up into the pulpit. But the Lord directed me to one, and, before I had preached half an hour, the Blessed Spirit began to move the hearers' hearts in a very awful manner. Young, and especially many old people were surprisingly affected. At dinner, I spoke with such vigour against sending unconverted ministers into the ministry, that two ministers with tears in their eyes, publicly confessed, that they had laid hands on two young men without so much as asking them, "whether they were born again of God, or not?" After dinner, I prayed, and one old minister was so deeply convicted, that calling Mr. Noble and me out, with great difficulty, (because of this weeping), he desired our prayers, "for," said he, "I have been a scholar, and have preached the doctrines of grace a long time, but I believe I have never felt the power of them in my own soul." Oh, that all unconverted ministers were brought to make the same confession! After having by prayer recommended him to God, I took horse, rejoicing exceedingly in spirit, to see how our Lord was getting Himself the victory, in a place where Mr. Davenport, a native of Stanford, a minister of the blessed Jesus, had been slighted and despised.

Here I think it proper to set up my Ebenezer, before I enter into the Province of New York, to give God thanks for sending me to New England. I have now had an opportunity of seeing the greatest and most populous part of it. On many accounts, it certainly excels all other provinces in America; and, for the establishment of religion, perhaps all other parts of the world. The towns all through Connecticut, and eastward toward York, in the Province of Massachusetts, near the river-side, are large and well peopled. Every five miles or perhaps less, you have a meeting-house; and, I believe, there is no such thing as a pluralist or non-resident minister in both provinces. Many, nay most that preach, I fear, do not experimentally know Christ; yet I cannot see much worldly advantage to tempt them to take up the sacred function. Few country ministers, I have been informed, have sufficient money allowed them to maintain a family. God has remarkably, at sundry times and in divers manners, poured out His Spirit in several parts; and it often refreshed my soul to hear of the faith of their good forefathers, who first settled in these parts. Notwithstanding they had their foibles, surely they were a set of righteous men. They certainly followed our Lord's rule, "by seeking first the Kingdom of God and His Righteousness," and all other things God added unto them. I think the ministers preaching almost universally by note, is a mark that they have, in a great measure, lost the old spirit of preaching. Though all are not to be condemned who use notes, yet it is a symptom of the decay of religion, when reading sermons becomes fashionable where *extempore* preaching did once almost universally prevail. When the spirit of prayer began to be lost, then forms of prayer were invented, and, I believe, the same observation will hold good as to preaching. The civil government of New England seems to be well regulated; and, I think, at the opening of all their courts, either the judge or a minister begins with a prayer. Family worship, I believe, is generally kept up: and the negroes are better used than in any other province I have yet seen. In short, I like New England exceeding well. Send forth, O Lord, Thy light and Thy truth, and for Thy infinite mercy's sake, shew Thou hast a peculiar delight in these habitable parts of the earth! Amen.

source 2

A Thousand People Hang on His Words

Sarah Pierrepont Edwards

> Mrs. Edwards's summing up of Whitefield, after his visit to North-ampton, is a gracious and ladylike piece of writing. It is also one of the most loaded descriptions of Whitefield; each attribute listed was controversial. One wonders what Mr. and Mrs. Edwards said about him in the privacy of their bedchamber.

October 24, 1740

DEAR BROTHER JAMES:

I want to prepare you for a visit from the Rev. Mr. Whitefield, the famous preacher of England. He has been sojourning with us, and, after visiting a few of the neighbouring towns, is going to New Haven, and from thence to New York. He is truly a remarkable man, and, during his visit, has, I think, verified all that we have heard of him. He makes less of the doctrines than our American preachers generally do, and aims more at affecting the heart. He is a born orator. You have already heard of his deep-toned, yet clear and melodious, voice. It is perfect music. It is wonderful to see what a spell he casts over an audience by proclaiming the simplest truths of the Bible. I

Sarah Pierrepont Edwards to James Pierrepont, October 24, 1740, Luke Tyerman, *The Life of the Rev. George Whitefield* (New York: Anson D. Randolph & Co., 1877), I, 428–29.

have seen upwards of a thousand people hang on his words with breathless silence, broken only by an occasional half-suppressed sob. He impresses the ignorant, and not less the educated and refined. It is reported that while the miners of England listened to him, the tears made white furrows down their smutty cheeks. So here, our mechanics shut up their shops, and the day-labourers throw down their tools, to go and hear him preach, and few return unaffected. A prejudiced person, I know, might say that this is all theatrical artifice and display; but not so will any one think who has seen and known him. He is a very devout and godly man, and his only aim seems to be to reach and influence men the best way. He speaks from a heart all aglow with love, and pours out a torrent of eloquence which is almost irresistible. I wish him success in his apostolic career; and, when he reaches New Haven, you will, I know, shew him warm hospitality.

Yours, in faithful affection,
SARAH.

source 3

Benjamin Franklin Describes
George Whitefield

Benjamin Franklin

> Wily Benjamin Franklin was one of very few American colonists who were not prejudiced in their attitude toward Whitefield, for most Americans either loved or despised him. Franklin's objectivity was based on the fact, as he said, that "We had no religious connection." The passage below from *The Autobiography* tells us as much about Franklin as it does about Whitefield.

In 1739 arrived among us from Ireland the Reverend Mr. Whitefield, who had made himself remarkable there as an itinerant preacher. He was at first permitted to preach in some of our churches; but the clergy, taking a dislike to him, soon refus'd him their pulpits, and he was oblig'd to preach in the fields. The multitudes of all sects and

Benjamin Franklin, *The Autobiography*, Albert Henry Smyth, ed., *The Writings of Benjamin Franklin . . . With a Life and Introduction* (New York: The Macmillan Company, 1905), I, 354–59.

denominations that attended his sermons were enormous, and it was matter of speculation to me, who was one of the number, to observe the extraordinary influence of his oratory on his hearers, and how much they admir'd and respected him, notwithstanding his common abuse of them, by assuring them they were naturally *half beasts and half devils.* It was wonderful to see the change soon made in the manners of our inhabitants. From being thoughtless or indifferent about religion, it seem'd as if all the world were growing religious, so that one could not walk thro' the town in an evening without hearing psalms sung in different families of every street.

And it being found inconvenient to assemble in the open air, subject to its inclemencies, the building of a house to meet in was no sooner propos'd, and persons appointed to receive contributions, but sufficient sums were soon receiv'd to procure the ground and erect the building, which was one hundred feet long and seventy broad, about the size of Westminster Hall; and the work was carried on with such spirit as to be finished in a much shorter time than could have been expected. Both house and ground were vested in trustees, expressly for the use of any preacher of any religious persuasion who might desire to say something to the people at Philadelphia; the design in building not being to accommodate any particular sect, but the inhabitants in general; so that even if the Mufti of Constantinople were to send a missionary to preach Mohammedanism to us, he would find a pulpit at his service.

Mr. Whitefield, in leaving us, went preaching all the way thro' the colonies to Georgia. The settlement of that province had lately been begun, but, instead of being made with hardy, industrious husbandmen, accustomed to labour, the only people fit for such an enterprise, it was with families of broken shop-keepers and other insolvent debtors, many of indolent and idle habits, taken out of the jails, who, being set down in the woods, unqualified for clearing land, and unable to endure the hardships of a new settlement, perished in numbers, leaving many helpless children unprovided for. The sight of their miserable situation inspir'd the benevolent heart of Mr. Whitefield with the idea of building an Orphan House there, in which they might be supported and educated. Returning northward, he preach'd up this charity, and made large collections, for his eloquence had a wonderful power over the hearts and purses of his hearers, of which I myself was an instance.

I did not disapprove of the design, but, as Georgia was then destitute of materials and workmen, and it was proposed to send them from Philadelphia at a great expense, I thought it would have been better to have built the house here, and brought the children to it. This I advis'd; but he was resolute in his first project, rejected my counsel, and I therefore refus'd to contribute. I happened soon after to attend one of his sermons, in the course of which I perceived he intended to finish with a collection, and I silently resolved he should get nothing from me. I had in my pocket a handful of copper money, three or four silver dollars, and five pistoles in gold. As he proceeded I began to soften, and concluded to give the coppers. Another stroke of his oratory made me asham'd of that, and determin'd me to give the silver; and he finish'd so admirably, that I empty'd my pocket wholly into the collector's dish, gold and all. At this sermon there was also one of our club, who, being of my sentiments respecting the building in Georgia, and suspecting a collection might be intended, had, by precaution, emptied his pockets before he came from home. Towards the conclusion of the discourse, however, he felt a strong desire to give, and apply'd to a neighbour, who stood near him, to borrow some money for the purpose. The application was unfortunately [made] to perhaps the only man in the company who had the firmness not to be affected by the preacher. His answer was, *"At any other time, Friend Hopkinson, I would lend to thee freely; but not now, for thee seems to be out of thy right senses."*

Some of Mr. Whitefield's enemies affected to suppose that he would apply these collections to his own private emolument; but I, who was intimately acquainted with him (being employed in printing his Sermons and Journals, etc.), never had the least suspicion of his integrity, but am to this day decidedly of opinion that he was in all his conduct a perfectly *honest man;* and methinks my testimony in his favour ought to have the more weight, as we had no religious connection. He us'd, indeed, sometimes to pray for my conversion, but never had the satisfaction of believing that his prayers were heard. Ours was a mere civil friendship, sincere on both sides, and lasted to his death.

The following instance will show something of the terms on which we stood. Upon one of his arrivals from England at Boston, he wrote to me that he should come soon to Philadelphia, but knew not where he could lodge when there, as he understood his old friend and host, Mr. Benezet was removed to Germantown. My answer was, "You know my house; if you can make shift with its scanty accommodations, you will be most heartily welcome." He reply'd, that if I made that kind offer for Christ's sake, I should not miss of a reward. And I returned, *"Don't let me be mistaken; it was not for Christ's sake, but for your sake."* One of our common acquaintance jocosely remark'd, that, knowing it to be the custom of the saints, when they received any favour, to shift the burden of the obligation from off their own shoulders, and place it in heaven, I had contriv'd to fix it on earth.

The last time I saw Mr. Whitefield was in London, when he consulted me about his Orphan House concern, and his purpose of appropriating it to the establishment of a college.

He had a loud and clear voice, and articulated his words and sentences so perfectly, that he might be heard and understood at a great distance, especially as his auditories, however numerous, observ'd the most exact silence. He preach'd one evening from the top of the Court-house steps, which are in the middle of Market-street, and on the west side of Second-street, which crosses it at right angles. Both streets were fill'd with his hearers to a considerable distance. Being among the hindmost in Market-street, I had the curiosity to learn how far he could be heard, by retiring backwards down the street towards the river; and I found his voice distinct till I came near Front-street, when some noise in that street obscur'd it. Imagining then a semicircle, of which my distance should be the radius, and that it were fill'd with auditors, to each of whom I allow'd two square feet, I computed that he might well be heard by more than thirty thousand. This reconcil'd me to the newspaper accounts of his having preach'd to twenty-five thousand people in the fields, and to the antient histories of generals haranguing whole armies, of which I had sometimes doubted.

By hearing him often, I came to distinguish easily between sermons newly compos'd, and those which he had often preach'd in the course of his travels. His delivery of the latter was so improv'd by frequent repetitions that every accent, every emphasis, every modulation of voice, was so perfectly well turn'd and well plac'd, that, without being interested in the subject, one could not help being pleas'd with the discourse; a pleasure of much the same kind with that receiv'd from an excellent piece of musick. This is an advantage itinerant preachers have over those who are stationary, as the latter can not well improve their delivery of a sermon by so many rehearsals.

His writing and printing from time to time gave great advantage to his enemies; unguarded expressions, and even erroneous opinions, delivered in preaching, might have been afterwards explain'd or quali-fi'd by supposing others that might have accompani'd them, or they might have been deny'd; but *litera scripta manet*. Critics attack'd his writings violently, and with so much appearance of reason as to dimin-ish the number of his votaries and prevent their encrease; so that I am of opinion if he had never written any thing, he would have left behind him a much more numerous and important sect, and his repu-tation might in that case have been still growing, even after his death, as there being nothing of his writing on which to found a censure and give him a lower character, his proselytes would be left at liberty to feign for him as great a variety of excellences as their enthusiastic ad-miration might wish him to have possessed.

source 4

What Shall We Do
to Be Saved?

Thomas Prince

Thomas Prince was a well-known and learned Congregational minister and historian in Boston. His support of the Awakening lent it
prestige and some dignity.

Boston, December 6, 1741.

REVEREND AND DEAR SIR,

I am now to inform you, that since my last our exalted Saviour
has been riding forth in his Magnificence and Glory thro' divers Parts
of our Land, in so triumphant a Manner as has never been seen or

Thomas Prince to George Whitefield, Boston, December 6, 1741, *The Glasgow-Weekly-History*, *1743*, reprinted in Massachusetts Historical So-ciety *Proceedings*, LIII (Boston, 1920), 204–5.

heard among us, or among any other People as we know of, since the Apostles Days. He is daily making his most resolute Opposers to fall down under him: And almost every Week we hear of new and sur- prising Conquests, and even almost all at once, and in a Manner over whole Congregations, where whole Assemblies lay as Congregations of the Dead; the Day of the Power of Christ comes at once upon us, and they are almost altogether, both Whites and Blacks, both Old and Young, both Prophane and Moral, awakened, and made alive to God. It is astonishing to see some who were like incarnate Devils, thrown at once into such extreme Distress as no Pen can possibly describe, or absent Mind imagine; and in two or three Days Time turn'd into eminent Saints, full of Divine Adoration, Love, and Joy unspeakable and full of Glory.

Amazing Works of this kind were a Fortnight ago begun, and are now going on at *Taunton* and *Middleborough,* about forty Miles South of *Boston,* at which last my Brother and Sister live, who so earnestly desired your Help; as also in *Bridgewater* about thirty Miles, and at *Abington* about twenty-three; and a little before at *York, Ipswich, Rowley, Cape Anne,* and thence to *Rittery* [Kittery] and *Berwick:* And on a Day of Fasting and Prayer at *Portsmouth, Dis- catoway* [Piscataqua], which you know is the Metropolis of *New- hampshire,* wherein both Mr. F.— and Mr. S—'s Congregation are join'd; before the Day was over, the Spirit of God came down and seiz'd them at once by Scores and Hundreds, that every one in the large Congregation clearly saw and own'd it was a Work of God, and in three Days time it was computed there were a Thousand in that Town in deep Distress about their Souls, and crying out in the Bitterness of Anguish, *What shall we do to be saved?* Yea, the won- drous Work even spread into the Church of *England* there: And this Evening I saw a Letter thence, wherein a Gentleman writes, that even Mr. B.—, the Church of *England* Clergyman there, who has often preach'd against this Work with distinguish'd Violence, has now declar'd he is convinc'd it is a wondrous Work of God.

source 5

The Terror of the Law

Jonathan Edwards

Luckily for his Northampton congregation, Jonathan Edwards did not preach only of hell-fire. To acquaint his hearers with "eternal death" was but part of his duty, for "conviction" was preparation for conversion and grace, which in turn required a wholly different set of ideas to affect the hearts of men. Nevertheless, the terrors of hell were part of the revivalist's technique, and Edwards was very impressive when he described them, as the following passage from one of his revival sermons illustrates.

INQUIRY. Some may be ready to say, If this be the case, if impenitent sinners can neither shun future punishment, nor deliver themselves from it, nor bear it; then what will become of them?

ANSWER. They will wholly sink down into eternal death. There will be that sinking of heart, of which we now cannot conceive. We see how it is with the body when in extreme pain. The nature of the body will support itself for a considerable time under very great pain, so as to keep from wholly sinking. There will be great struggles, lamentable groans and panting, and it may be convulsions. These are the strugglings of nature to support itself under the extremity of the pain. There is, as it were, a great lothness in nature to yield to it; it cannot bear wholly to sink.

But yet sometimes pain of body is so very extreme and exquisite, that the nature of the body cannot support itself under it; however loth it may be to sink, yet it cannot bear the pain; there are a few struggles, and throes, and pantings, and it may be a shriek or two, and then nature yields to the violence of the torments, sinks down, and the body dies. This is the death of the body. So it will be with the soul in hell; it will have no strength or power to deliver itself; and its torment and horror will be so great, so mighty, so vastly disproportioned to its strength, that having no strength in the least to support itself,

Jonathan Edwards, "The Future Punishment of the Wicked Unavoidable and Intolerable," *The Works of President Edwards* (New York: Leavitt and Co., 1851), IV, 260–61.

although it be infinitely contrary to the nature and inclination of the soul utterly to sink; yet it will sink, it will utterly and totally sink, without the least degree of remaining comfort, or strength, or courage, or hope. And though it will never be annihilated, its being and perception will never be abolished; yet such will be the infinite depth of gloominess that it will sink into, that it will be in a state of death, eternal death.

The nature of man desires happiness; it is the nature of the soul to crave and thirst after well-being; and if it be under misery, it eagerly pants after relief; and the greater the misery is, the more eagerly doth it struggle for help. But if all relief be withholden, all strength overborne, all support utterly gone; then it sinks into the darkness of death.

We can conceive but little of the matter; we cannot conceive what that sinking of the soul in such a case is. But to help your conception, imagine yourself to be cast into a fiery oven, all of a glowing heat, or into the midst of a glowing brick-kiln, or of a great furnace, where your pain would be as much greater than that occasioned by accidentally touching a coal of fire, as the heat is greater. Imagine also that your body were to lie there for a quarter of an hour, full of fire, as full within and without as a bright coal of fire, all the while full of quick sense; what horror would you feel at the entrance of such a furnace! And how long would that quarter of an hour seem to you! If it were to be measured by a glass, how long would the glass seem to be running! And after you had endured it for one minute, how overbearing would it be to you to think that you had it to endure the other fourteen!

But what would be the effect on your soul, if you knew you must lie there enduring that torment to the full for twenty-four hours! And how much greater would be the effect, if you knew you must endure it for a whole year; and how vastly greater still, if you knew you must endure it for a thousand years! O then, how would your heart sink, if you thought, if you knew, that you must bear it forever and ever! That there would be no end! That after millions of millions of ages, your torment would be no nearer to an end, than ever it was; and that you never, never should be delivered!

But your torment in hell will be immensely greater than this illustration represents. How then will the heart of a poor creature sink under it! How utterly inexpressible and inconceivable must the sinking of the soul be in such a case!

This is the death threatened in the law. This is dying in the highest sense of the word. This is to die sensibly; to die and know it; to be sensible of the gloom of death. This is to be undone; this is worthy of the name of destruction. This sinking of the soul under an infinite weight, which it cannot bear, is the gloom of hell. We read in Scripture of the blackness of darkness; this is it, this is the very thing. We read in Scripture of sinners being lost, and of their losing their souls: this is the thing intended; this is to lose the soul: they that are the subjects of this are utterly lost.

source 6

The Conversion of
Isaac Backus and His Call
to the Ministry

Isaac Backus

The Awakening caught up with Isaac Backus of Norwich, Connecticut, in the summer of 1741 when he was seventeen years old. His conversion was followed by a good deal of experience as a separatist itinerant and exhorter and then, once he was in Middleborough, Massachusetts, as a Baptist. A dissenter in a Congregational colony, Backus lived out his long life as New England's strongest advocate of the separation of church and state. An account of his conversion and his reasons for becoming a minister are simply and honestly described in these two passages from the Backus manuscripts.

A. AS I WAS MOWING IN THE FIELD ALONE

Although I was often warned and Exorted (especially by my godly mother), To fly from the Wrath to come:—yet I Never was under any

Isaac Backus his writing Containing Some Particular account of my Conversion, Isaac Backus Papers, Andover-Newton Theological School Library, Newton Center, Massachusetts. For clarity's sake the editor has extended numerous abbreviations and symbols which appeared in the manuscript of this document.

Powerful Conviction Till the year 1741. When it pleased the Lord to cause a very general awakening Thro' the Land; especially in Norwich The work was so remarkable that the Children & yong People were broke off from Their Plays & frolicks (which they were much ingaged in before) So that if one Went from one end of the Town Street to the Other they would scarcely see a Child at Play:—& the yong People left off their Frolicks for Several years. Many were Hopefully Converted & many others under Powerful Conviction & a general thoughtfulness seemed to appear on the minds of people[.] Before these times I never thought my Self in a Safe State but yet eased my Self with Purposes of turning by & by when I should have a more convenient Season.

But now in the begining of this awakening in May & June 1741, God by his Spirit was Pleased in Infinite Mercy to bring eternal Things near to my Soul and to shew me the Dredful danger of delays.—& from the first Begining of this work my mind was Deeply Impressed with this Persuasion That now God was Come with the offers of his grace to my Soul & this was The Last Season that ever I should Have to make my Peace with him Which made my Soul to awake & strive In good earnest for Salvation. But alas! I Was awfully Ignorant both of my own heart And of the way of Life by Christ. However I know no other way but only to go and Hear the most Powerfull preaching that I could And I hoped that by & by I should be converted—now about this time it pleased The Lord to Send many Powerfull Preachers To Norwich.—the first that Preacht There was Mr. Wheelock June 2; & in the latter end of the same week I heard mr Jedidiah Mills also Preach there two Sermons. I was something Affected. But I couldn't for my life get Such Convictions as I wanted & as I thought many others had.—after this in the Summer I had oppertunity to hear Sundry awakening discourses but my case remained much The Same.—About the begining of August Mr. Devenport Came to Norwich & preached There three days going in an exceeding Earnest & Powerful manner; & I apprenhend that his labours were that most blest For my Conversion of any one mans—

Mr Pomroy & Wheelock met him Next at Norwich & we had a great Many powerfull meetings while they Were here: and a number I Believe were converted but I seemed to be left: the thoughts of which seemed dreadfull to my Soul.—

Nothing now distresed me more Than to find that hearing the most Powerful preaching & also the Sreaks & Cries of Souls under Concern—did not Affect me as I desired But my heart Felt hard notwithstanding.—But in Truth the Lord was then letting me in to see Something of the Plague of my Heart & the fountain of Corruption that was there. I Remember once in Particular that I had such a view of my Heart that I really Saw that there want a Sin In the whole World but I had the seeds of it In my heart:—this Brought on distress Indeed tho' not such as I was seeking after[.] No;—my tears & good desires that I was Seeking for were dried up & instead of a Penitent frame I found a hard heart & a Blind mind & Instead of good desires I found Dreadful Qurelings against God, especially Against his Sovereingnty & the freeness of his Grace—that he was no ways obliged to give me His Grace, let me do as much as I would—

But then again to see that I had Such Corruptions in me brought on fresh distress That I knew not which way to turn.

Sometimes I thought that my Convictions Were weareing off & that Gods Spirit would Leave me but that appeared dredfull indeed, for I Still thought that this was the last Call That I ever should have—thus I worried Along for Some weeks—one Sabbath as our old minister was speaking to persons under Convictions he laid out a case much Like mine & then said "if this be your Case don't be discouraged, but see if God don't Speadily appear for your help! ["] or words To that Purpose.—Immediately the Tempter clapt in "there says he is Incouragement for you to be easey & not Trouble your Self so much; you may be Converted by & by, ["] but I was made so Sensable that it was a Temptation that I was filled with more distress. . . .

Not Long after this on August the 29 1741 as I was mowing in the field alone—I was thinking of my case; & all my past Life seemed to be brought fresh to my view And it appeared indeed nothing but a life of Sin—I felt so that I left work & went & sat Down under a shadey tree; & I was brought to Look Paticularly into my duties & striveings How I had tryed to get help by awakening Preaching but found it fail:—had tried to Mend my Self by my Tears prayers & Promises of doing better but all in vain—my heart was Hard & full of corruption still and it Appeared clear to me then that I had tryed Every way that Posibly I Could & if I perished Forever I could do no more—& the Justice of God Shined so clear Before my eyes in Condemning such a guilty Rebel that I could say no more—but fell at his feet[.] I see that I was in his hands & he had a right To do with me just as he Pleased And I lay like a Dead Vile Creature before him. I felt a calm in my mind—them tossings And tumults that I felt before seemed to be gone.

And Just in that Critical moment God who caused the light to Shine out of Darkness,—Shined into my heart with such A discovery of that glorious Righteousness Which fully Satesfies the Law that I had Broken; & of the Infinite fullness that there Is in Christ to Satesfie the wants of just Such A helpless Creature as I was & these Blessings Were held forth so freely to my Soul—That my Whole Heart was atracted and Drawn away after God & Swallowed up with Admiration in viewing his Divine glories.

Never did his Word appear So before as It did now:—it appeared So glorious & Such Infallible Truth that I could with the greatest Freedom Rest my Eternal all upon what God hath Spoken—now the way of Salvation appeared so excellent & glorious That I Wondered that I had stood out So long against Such a Blessed Redeemer Yea I wondered that all the World didn't Come to him—

And now my Burden (that was so dreadful Heavey before) was gone: that tormenting fear That I had was taken away & I felt a sweet Peace & rejoiceing in my soul.—But yet all This time I hadn't one thought that this was That Which is Called Conversion; it was so Different from the notion that I had of it before

Yet afterward when I came to think of my Case I couldn't get that distress which I had Before. But I thought I should be converted Hereafter: thus I went on for some days

—The first time that I thought I Was Converted was 2 or 3 days after at an Evening meeting when they read Mr. Whitefields Sermon on Acts 19:2—Where he lays down Sundry marks of trial To know whether we are saints or not. I thought When they began to Read that, it want so Particularly for me for I never thought my Self Converted; & so I didn't need to try That point.—But as they read along ere I was aware my Soul gave in that I had Felt what he there Lays down as marks of true grace:—as particalarly a spirit of Prayer—A lothing & hatred of Sin—Love to the Bretherin &c. and then The Lord was Pleased to give me some Sweet Sealings of the holy Spirit of Promise: & a comfortable Season it was to my Soul.—O Bless the Lord My Soul! & forget not all his benefits

These mercies of the Lord are Greater than Tongue can express—
that He should deliver my Soul from going down To the Pit & from
the Clutches of the old Dragon.—And that in the forepart of my
Days.—Being then 17n years & 7 months Old[.] O! for Strength always to live to The glory of God *Amen!*————

B. HE CALLED ME TO PREACH HIS GOSPEL

Hitherto a private life had been my choice and delight, but a new
scene was before me, which I had no idea of, until I was lead into it in
the following manner. Mr. Solomon Paine was ordained at Canterbury,
September 10, and Mr. Thomas Stevens at Plainfield, September 11,
where I saw many wonderful things; and as I came home the next day,
a conviction seized my mind, that God had given me abilities that his
church had a right to the use of, and that I could not withold it with
a clear conscience. And soon after in the dead of the night, a spirit of
prayer for divine teaching was given me in a remarkable manner, and
eternal things were brought into a near view, with a clear sight of the
truth and harmony of the holy scriptures, with that word upon it, "Son
of man eat the roll," and my bowels were filled with it, as sensably as
I ever eat temporal food. Yet I had no conclusion then that I ever should
preach the gospel in public. But a few days after our minister invited
me to go with him to Colchester and Lyme, where was a revival of religion, and I went accordingly, and two souls were hopefully converted
in the journey, and I returned home with rejoicing and much life in my
soul. And the day after which was September 27, 1746, new vews were
given me, beyond what I had before. My business lead me out to work
alone in the woods, where I had none to interrup me, in such a converse
with my God as I never had before. His former teaching now came to
this point, that he called me to preach his gospel; and I was lead to
count the cost as distinctly as ever I cast up any temporal sum. Many
and great enemies appeared in my way, as reproaches, losses, imprison-
ment and death; but God shewed me that he could make them all to
fly out of the way, as easily as the chaff flies before the wind. Then

An Account of the Life of Isaac Backus, pp. 31–39, 43, Isaac Backus
Papers, Andover-Newton Theological School Library, Newton Center,
Massachusetts.

came up my own ignorance and weakness; but he gave me to see that he had knowledge and strength for me. Upon this I pleaded that I was slow of speech, and very bashful; but he said, Cannot he who formed man's mouth make him speak? This I could not deny, yet I pleaded that if I should go & preach the gospel, and should be successful therein, I might be lifted up with pride, and fall as many others had done. This looked like a great mountain vastly abve all other difficulties; but God said, *My grace is sufficient for thee;* which took the mountain out of the way, and every excuse was gone, even so that it appeared to be triffling with Divine majesty to attempt to make another objection. Therefore I was enabled there to give up my soul a body afresh to God, with all my interest, to go & serve him in preaching his gospel, by his direction and assistance.

He then gave me a particular message, from the fifty third psalm, to lay open the universal corruption of mankind. And as our church allowed each brother free liberty to improve his gift in teaching, as God gave it to him, I delivered that message the next day, which was the Lords day, with special clearness, and to the acceptance of the church. And as I was then free from all worldly engagements, I devoted my whole time to that great work. And I went with Mr. Hide, our minister, to Preston, Stonington and Westerly, where we saw much of the power of the gospel upon the souls of men; and we returned home on October 8. Mr John Fuller was then keeping school in Norwich, who had been internally called to preach before, but had not obeyed it, and thereby had brought great darkness upon his mind. He being at a meeting on October 10, a word took such hold of his mind as caused him to cry out, and to confess his former disobedience, and to give a moving exhortation; and he then devoted himself to the great work of preaching the gospel through the land.

.   .   .

A few men there came out against me, and some of them pretended to be perfect and immortal, as others have done since in our land. . . .

The Danger of
an Unconverted Ministry

Gilbert Tennent

Along with Jonathan Edwards's *Sinners in the Hands of an Angry God,* Gilbert Tennent's Nottingham Sermon was one of the most memorable contributions to the Great Awakening. Whereas Edwards's sermon frightened only sinners, Tennent's sermon frightened, offended, and exasperated a number of ministers, particularly those in his own Philadelphia Synod who were not convinced of the genuineness of the Revival.

2. The Ministry of natural Men is uncomfortable to gracious Souls.

The Enmity that is put between the Seed of the Woman and the Seed of the Serpent, will now and then be creating Jarrs: And no wonder; for as it was of old, so it is now, *He that was born after the Flesh, persecuted him that was born after the Spirit.* This Enmity is not one Grain less, in unconverted Ministers, than in others; tho' possibly it may be better polished with Wit and Rhetorick, and gilded with the specious Names of Zeal, Fidelity, Peace, good Order, and Unity.

Natural Men, not having true Love to Christ and the Souls of their Fellow-Creatures, hence their Discourses are cold and sapless, and as it were freeze between their Lips! And not being sent of GOD, they want that divine Authority, with which the faithful Ambassadors of Christ are clothed, who herein resemble their blessed Master, of whom it is said, That *He taught as one having Authority, and not as the Scribes.* Matth. 7. 29.

Gilbert Tennent, *The Danger of an Unconverted Ministry, Considered in a Sermon . . . Preached at Nottingham, in Pennsylvania, March 8. Anno 1739, 40.* (Philadelphia: Printed by Benjamin Franklin in Market-street, 1740), pp. 8–9, 10–11, 13–14, 18–19, 20–22, 27, 30–31.

. . . All the Doings of unconverted Men, not proceeding from the Principles of Faith, Love, and a new Nature, nor being directed to the divine Glory as their highest End, but flowing from, and tending to Self, as their Principle and End; are doubtless damnably wicked in their Manner of Performance, and do deserve the Wrath and Curse of a Sin-avenging GOD; neither can any other Encouragement be justly given them, but this, That in the Way of Duty, there is a Peradventure or Probability of obtaining Mercy.

And Natural Men, wanting the Experience of those spiritual Difficulties, which pious Souls are exposed to, in this Vale of Tears; they know not how to speak a Word to the Weary in Season.

Their Prayers are also cold; little Child-like Love to God, or Pity to poor perishing Souls, runs thro' their Veins.

Their Conversation hath nothing of the Savour of Christ, neither is it perfumed with the Spices of Heaven. They seem to make as little Distinction in their Practice, as Preaching. They love these Unbelievers, that are kind to them, better than many Christians, and chuse them for Companions; contrary to *Ps.* 15. 4. *Ps.* 119. 115. & *Gal.* 6. 10. Poor Christians are stunted and starv'd, who are put to feed on such bare Pastures, and such dry Nurses. . . . It's only when the wise Virgins sleep, that they can bear with those dead Dogs, that can't bark; that when the LORD revives his People, they can't but abhor them! O! it is ready to break their very Hearts with Grief, to see, how lukewarm these Pharisee-Teachers are in their publick Discourses, while Sinners are sinking into Damnation, in Multitudes! . . .

. . .

4. The Ministry of natural Men is dangerous, both in respect of the Doctrines, and Practice of Piety. The Doctrines of *Original Sin*, *Justification by Faith alone*, and the other Points of *Calvinism*, are very cross to the Grain of unrenewed Nature. And tho' Men, by the Influence of a good Education, and Hopes of Preferment, may have the Edge of their natural Enmity against them blunted; yet it's far from being broken or removed; it's only the saving Grace of God, that can give us a true Relish, for those Nature-humbling Doctrines; and so effectually secure us from being infected by the contrary. Is not the Car-

nality of the Ministry, one great Cause of the general Spread of *Arminianism, Socinianism, Arianism,* and *Deism,* at this Day through the World?

And alas! what poor Guides are natural Ministers to those, who are under spiritual Trouble? they either slight such Distress altogether, and call it Melancholy, or Madness, or daub those that are under it, with untemper'd Mortar. Our LORD assures us, That the Salt which hath lost its Savour, is good for nothing; some say, 'It genders Worms and Vermine.' Now, what Savour have Pharisee-Ministers? In Truth, a very stinking One, both in the Nostrils of God and good Men. 'Be these Moral Negroes never so white in the Mouth (as one Expresseth it), yet will they hinder, instead of helping others, in at the strait Gate.' Hence is that Threatning of our LORD, against them, *Mat.* 23. 13. *Wo unto you Scribes and Pharisees, Hypocrites; for ye shut up the Kingdom of Heaven against Men; for ye neither go in yourselves, nor suffer those that are entering, to go in.* Pharisee-Teachers will with the utmost Hate oppose the very Work of God's Spirit, upon the Souls of Men; and labour by all Means to blacken it, as well as the Instruments, which the Almighty improves to promote the same; if it comes near their Borders, and interferes with their Credit or Interest. Thus did the Pharisees deal with our Saviour.

:●:   :●:   :●:

2. From what has been said, we may learn, That such who are contented under a dead Ministry, have not in them the Temper of that Saviour they profess. It's an awful Sign, that they are as blind as Moles, and as dead as Stones, without any spiritual Taste and Relish. And alas! isn't this the Case of Multitudes? If they can get one, that has the Name of a Minister, with a Band, and a Black Coat or Gown to carry on a Sabbath-days among them, although never so coldly, and insuccessfully; if he is free from gross Crimes in Practice, and takes good Care to keep at a due Distance from their Consciences, and is never troubled about his Insuccessfulness; O! think the poor Fools, that is a fine Man indeed; our Minister is a prudent charitable Man, he is not always harping upon Terror, and sounding Damnation in our Ears, like some rash-headed Preachers, who by their uncharitable

Methods, are ready to put poor People out of their Wits, or to run them into Despair; O! how terrible a Thing is that Despair! Ay, our Minister, honest Man, gives us good Caution against it. Poor silly Souls! consider seriously these Passages, of the Prophet *Jeremiah*, c. 5. 30, 31. . . .

• • •

4. If the Ministry of natural Men be as it has been represented; Then it is both lawful and expedient to go from them to hear Godly Persons; yea, it's so far from being sinful to do this, that one who lives under a pious Minister of lesser Gifts, after having honestly endeavour'd to get Benefit by his Ministry, and yet gets little or none, but doth find real Benefit and more Benefit elsewhere; I say, he may lawfully go, and that frequently, where he gets most Good to his precious Soul, after regular Application to the Pastor where he lives, for his Consent, and proposing the Reasons thereof; when this is done in the Spirit of Love and Meekness, without Contempt of any, as also without rash Anger, or vain Curiosity.

• • •

If God's People have a Right to the Gifts of all God's Ministers, pray, why mayn't they use them, as they have Opportunity? And if they should go a few Miles farther than ordinary, to enjoy those, which they profit most by; who do they wrong? . . .

• • •

To bind Men to a particular Minister, against their Judgment and Inclinations, when they are more edified elsewhere, is carnal with a Witness; a cruel Oppression of tender Consciences, a Compelling of Men to Sin: For he that doubts, is damn'd if he eat; and whatsoever is not of Faith, is Sin. . . .

If the great Ends of Hearing may be attained as well, and better, by Hearing of another Minister than our own; then I see not, why we should be under a fatal Necessity of Hearing him, I mean our Parish-Minister, perpetually, or generally. Now, what are, or ought to be, the

Ends of Hearing, but the Getting of Grace, and Growing in it? *Rom.*
10. 14. 1 *Pet.* 2. 2. *As Babes desire the sincere Milk of the Word, that
ye may grow thereby.* (Poor Babes like not dry Breasts, and living Men
like not dead Pools.) Well then, and may not these Ends be obtained
out of our Parish-line? *Faith* is said to come by *Hearing*, Rom. 10. But
the Apostle doesn't add, *Your Parish-Minister.* Isn't the same Word
preach'd out of our Parish? and is there any Restriction in the Promises
of blessing the Word to those only, who keep within their Parish-line
ordinarily? If there be, I have not yet met with it; yea, I can affirm, that
so far as Knowledge can be had in such Cases, I have known Persons
to get saving Good to their Souls, by Hearing over their Parish-line;
and this makes me earnest in Defence of it.

That which ought to be the main Motive of Hearing any, *viz.* our
Souls Good, or greater Good, will excite us, if we regard our own eternal
Interest, to hear there, where we attain it; and he that hears with less
Views, acts like a Fool, and a Hypocrite.

Now, if it be lawful to withdraw from the Ministry of a pious
Man, in the Case aforesaid; how much more, from the Ministry of a
natural Man? Surely, it is both lawful and expedient, for the Reasons
offered in the Doctrinal Part of this Discourse: To which let me add a
few Words more.

To trust the Care of our Souls to those who have little or no Care
for their own, to those who are both unskilful and unfaithful, is contrary to the common Practice of considerate Mankind, relating to the
Affairs of their Bodies and Estates; and would signify, that we set light
by our Souls, and did not care what became of them. For if the Blind
lead the Blind, will they not both fall into the Ditch?

• • •

Again it may be objected, That the aforesaid Practice tends to
grieve our Parish-Minister, and to break Congregations in Pieces.

I answer, If our Parish-Minister be grieved at our greater Good,
or prefers his Credit before it; then he has good Cause to grieve over his
own Rottenness and Hypocrisie. And as for Breaking of Congregations
to Pieces, upon the Account of People's Going from Place to Place, to
hear the Word, with a View to get a greater Good; that spiritual Blindness and Death, that so generally prevails, will put this out of Danger.

It is but a very few, that have got any spiritual Relish; the most will venture their Souls with any Formalist, and be well satisfied with the sapless Discourses of such dead Drones.

. . .

And let those who live under the Ministry of dead Men, whether they have got the Form of Religion or not, repair to the Living, where they may be edified. Let who will, oppose it. . . . But tho' your Neighbours growl against you, and reproach you for doing your Duty, in seeking your Souls Good; bear their unjust Censures with Christian Meekness, and persevere; as knowing that Suffering is the Lot of Christ's Followers, and that spirit-Benefits do infinitely overbalance all temporal Difficulties.

And O! that vacant Congregations would take due Care in the Choice of their Ministers! Here indeed they should hasten slowly. The Church of *Ephesus* is commended, for Trying them which said they were Apostles, and were not; and for finding them Liars. Hypocrites are against all Knowing of others, and Judging, in order to hide their own Filthiness; like Thieves they flee a Search, because of their stolen Goods. But the more they endeavour to hide, the more they expose their Shame. Does not the spiritual Man judge all Things? Tho' he cannot know the States of subtil Hypocrites infallibly; yet may he not give a near Guess, who are the Sons of *Sceva*, by their Manner of Praying, Preaching, and Living? Many Pharisee-Teachers have got a long fine String of Prayer by Heart, so that they are never at a Loss about it; their Prayers and Preachings are generally of a Length, and both as dead as a Stone, and without all Savour. I beseech you, my dear Brethren, to consider, That there is no Probability of your getting Good, by the Ministry of Pharisees. For they are no Shepherds (no faithful ones) in Christ's Account. They are as good as none, nay, worse than none, upon some Accounts. For take them first and last, and they generally do more Hurt than Good. They serve to keep better out of the Places where they live; nay, when the Life of Piety comes near their Quarters, they rise up in Arms against it, consult, contrive and combine in their Conclaves against it, as a common Enemy, that discovers and condemns their Craft and Hypocrisie. And with what Art, Rhetorick, and Appearances of Piety, will they varnish their Opposition of Christ's Kingdom? As the Magicians imitated the Works of *Moses,* so do false Apostles, and deceitful Workers, the Apostles of Christ. **. . .**

source 8

The Awakening Reaches Backcountry Virginia

Patrick Henry, Sr.

> When the Awakening came to Virginia, it took the form of New Side Presbyterianism, brought there by some of Gilbert Tennent's "red-hot Bretheren." According to the Anglican clergy, the "Enthusiasts" were a threat to the Established Church. Patrick Henry, Sr., uncle of the Revolutionary patriot, was Rector of St. Paul's Parish in Hanover. His letter to the Bishop of London's Commissary in Virginia gives a point-by-point explanation of the Awakening's principles.

St. Pauls parish Hanover Feb. 13th, 1744/5

REVEREND SIR

I would have wrote you before now concerning the new Preachers that have lately seduc'd some unwary people in this Parish, had I not expected to be more distinctly inform'd of some of their principles and practices which I thought might render my account of them or their followers more full and satisfactory which please take as follows. There is in Pennsylvania a Synod of Protestant Dissenters consisting of about 40 members, one of whom viz Mr. John Thomson came to a certain Gentleman's house in our parish, on thursday the first of this month, intending to preach the Sunday following in the meeting house lately

The Reverend Patrick Henry, Sr., to Commissary William Dawson, Hanover, Va., February 13, 1745, *The William and Mary Quarterly,* series 2, I (1921), 261–66.

erected here, but when he with a few that accompany'd him, came to the house on Sunday morning, the followers of Robinson, Blair & Roan (whom I mentioned to you when at Wmsburg) shut the doors against him alledging he was an opposer of these three, the last of whom had wrote to some of them, requesting them in the name of the Lord, and for the Sake of Christ Jesus, not to allow Mr. Thomson to preach in their house, because he is an enemy to Christ & true religion. On hearing of this difference among them, I sent and invited Thomson to my house. He entertained me with a distinct account of these new light men, their peculiar tenets, and practices, their rise and progress to this time. He is, in my opinion, a man of learning and good Sense, a strenuous opposer of these new Preachers and Whitfield, having published two small treatises against them (which I think are very well performed) and I believe he is a man of piety and veracity. So that his information may be look'd upon as true. The substance of which with what I have upon other undoubted [?] is as follows. There is one Gilbert Tennent lately a leading man in the Synod of Presbyterians in Pennsylvania, who, with one Mr. Freelenhauson a Dutch Minister of Staten island, had several years before Mr. Whitfield appear'd in America broach'd some strange notions about religious matters, which some other younger Preachers imbibd from them, but they had not authority enough to impose these notions upon the people, till Whitfield coming over joind them, and then their notions and opinions were every where publishd, and being espoused by Whitfield and his followers, became the current Doctrines of that joint party; and at a meeting of the above mentioned Synod at Philadelphia in May 1741 this Tennent and eight more of the members openly declared their separation from the Synod, and have ever since that time continued to meet by themselves, to [?] a discipline of their own framing, and have ordaind a good many young Preachers, whom they send into all parts of America to disturb the established Churches of all denominations, requiring almost no other qualification in Candidates for Orders, than, what they call experiences of a work of grace in their hearts; and the Preachers who lately came into Hanover were three of those ordained by these Separatists above mentioned. The new doctrines these Schismaticks are at great pains to propagate and which their Missionaries publickly taught among us here were chiefly these following viz.

That antecedent to the very first beginning of a work of grace, there is a necessity of what they call, a Law work or common convictions, whereby the Sinner must be brought to despair, by way of preparation for Gospel grace, and some of them assert, That men must be willing to be damm'd, before they can obtain an interest in saving grace or mercy. And Roan who preachd in Hanover about Christmas last asserted in one of his publick discourses (as I was informd by one who heard him) That a Sinner, before he can be thoroughly converted, must experience this Law work in such a degree as to disbelieve the very being of a God. II. That every true Convert is able to give an historical narrative of the time and manner of his or her conversion. III. That every converted person is as assuredly sensible of the Spirit of God working in him, as he would be of a wound or stab, or any thing else that he knows by his outward senses. IV. That all true believers, and especially converted ministers have the spirit of discerning whereby they can distinguish a hypocrite or a formal professor, from a sincere Christian. And this Spirit is claimd by some here in Hanover, particularly Samuel Morris and Thomas Green two of my neighbours. V. That a true Christian may know whether a Minister be converted or not by hearing him preach or pray. This wild notion prevails among our Enthusiasts here, and I have been condemn'd by some of them as a stranger to true religion, & what they call the work of God, particularly by one Roger Shackleford who having come to Church last Sunday, in his way home told those about him, that I had preach'd Damnable doctrine, and he pitied me as being an unconverted graceless man. And now that I have mentiond Shackleford, I cannot omit informing you of another piece of his conduct. I sent him one of the Bps of London's letters for his perusal, and before he had read it half over, he returnd it to the person by whom I sent it, and told her that he was sure the Bishop was an unconverted man, and said he wished God would open his eyes to see the truth. VI. That a Minister being unconverted hath no call or authority from God to preach the Gospel and such a Minister's preaching, tho' he preach sound doctrine, can be of no saving use to the hearers. And thus by their pretended Spirit of discerning they apply the sentence of Condemnation to all ministers who are not of their way, and persuade as many as they can, to forsake their own Pastors as carnal graceless wretches, tho men of good principles and blameless lives. VII. That a regular ordination of a man to the holy Ministry, after due tryal and examination, is not the call of God, but of men only, the call of God with them being wholly inward by the Spirit and that therefore none ought to be admitted into the Ministry, but such as are sure of their conversion. VIII. That Christians are not obliged to adhere to their own respective Pastors, but ought to go to hear the word preachd where they think they receive the greatest benefit, or where they meet with the greatest gifts in the Preachers.

IX. They make little or no account of a sound profession of Doctrine, joind with a regular Christian conversation, as a ground of judging charitably concerning a man's gracious State, unless one can give a narrative of the work of the Spirit of God in his heart, to judge charitably of a man's state on any other account is called by them a murdering, barbarous charity.

X. They claim a right to examine whom they please concerning their spiritual state, and take them to pronounce such as dont please them in their answers, to be in a carnal damnd condition (These are their own words) This right to examine is common to both Preachers and people. XI. Both Preachers and people are great boasters of their assurance of salvation. They are so full of it here that the greatest number of those who have lately left the Church, and followed those Enthusiastik Preachers, last, as if they were there already; nay some people here who have always been justly reputed guilty of several immoralities such as do confidently assert that they are as sure of going to Heaven at cheating, lying, and even theft, and whose practices (I well know) are the same now as before, these very men do boast as much of their assurances, as others who are reckond blameless in their conversation: where such as these are so confident or rather impudent, you'l be less surpriz'd at what follows, viz, That their Preachers publickly tell their hearers, that they shall stand at the right hand of Christ in the day of Judgment, and condemn all of them who do not come to him at their call.

Having given you an abstract of their doctrines, I beg leave to add a few sentences relating to their practice especially that of the three Enthusiasts that preach'd lately in this Parish. These have been at great pains to vilifie the Clergy of this Colony and have told their followers, both in publick & private that they can never reap any benefit by going to hear them, because they are not the Servants of God, and have no authority to meddle with Holy things; They endeavour to give them a mean opinion of our Liturgy, but this I believe they have done chiefly in private, for I did not hear that they spoke against it in their Sermons, however I know, that their adherents generally disperse it and one of them (Thomas Green), told one of his Neighbours that it containd abundance of lies, and mentioned that sentence in the Te Deum (All the earth doth worship thee) as one. These three that were with us, as well as their brethren elsewhere, strive with all their might, to raise in their hearers, what they call convictions, which is thus performd. They thunder out [?] words and new coind phrases what they call the terrors of the law, [?] & scolding, calling the old people, Grey headed Devils and all promiscuously Damn'd, double damn'd whose souls are in hell, though they are alive on earth, Lumps of hellfire, incarnate Devils, 1000 times worse than Devils &c and all the while the Preacher exalts his voice puts himself into a violent agitation stamping & beating his Desk unmercifully until the weaker sort of his hearers being scar'd, cry out fall down & work like people in convulsion fits to the amazement of Spectators, and if a few only are thus brought down, the Preacher gets into a violent passion again, Calling out Will no more of you come to Christ? thundering out as before, till he has brought a quatum sufficit of his congregation to this condition and these things are extolld by the Preachers as the mighty power of God's grace in their hearts, and they who thus cry out and fall down are caressd and commended as the only penitent Souls who come to Christ, whilst they who don't, are often condemn'd by the lump as hardned wretches almost beyond the reach of mercy, insomuch that some who are not so season'd, impute it to the hardness of their own heart, and wish and pray to be in the like condition.

You may probably think, Sir, that I am a little hyperbolical in this last relation, but I beg leave to assure you, that I have unquestionable authority for the truth of it, and that they have acted in this parish in the same manner as I have now describd.

I am told that there are two or three of these Enthusiastic Preachers expected in Hanover next month, to administer the Sacrament of the Lord's Supper; I wish they could be prevented, or, at least be oblig'd to show their credentials, for they may be Jesuits for anything we know.

You have here inclosed some notes of a sermon preachd by the last of these Missionaries; I was to have transcribd it but have not been at leisure to do it. I purpose to wait on you at Wmsburg—as soon as my parochial & other business will allow, that I may have some further directions about my conduct with respect to these wild & wicked men, and am very respectfully Reverend Sir

Your most obedient humble Servant

PATRICK HENRY

Pray Sir, excuse some interlining
&c I being necessarily in hast.

part two

A SPIRIT OF
SUPERSTITION
AND ENTHUSIASM

New England and the French Prophets

Anonymous

> In the eighteenth century it was common to identify one's opponents with an unpopular cause or set of ideas. In 1742 there appeared in Boston a long tract called *The Wonderful Narrative* —formerly attributed to Charles Chauncy but recently questioned— which described the notorious French Prophets, an extreme group of Huguenots who had flourished earlier in the century. So that no one would miss the point, the author identified the activities of the Awakening with the enthusiasm of the Prophets in an Appendix, a part of which is reprinted here.

APPENDIX

• • •

Who are the Persons who see VISIONS and fall into TRANCES, and make Pretences to the *Spirit* in an *extraordinary* Manner? Look back into the History of former Days, and you will presently find who they are; not the *sober* and *judicious* among Christians, those who place Religion in that which is the *Life* and *Essence* of it: No, but the *weak* and *ignorant,* or those who are naturally of a *warm Imagination. Visions* and *Revelations* are to be met with *chiefly,* if not only, among those who have been esteemed *Enthusiasts,* and have proved themselves to be so: They were common among the *Montanists,* but not among the other Christians in that Day; among the FRENCH PROPHETS, and those in *that Way* in *England,* but not among the Christians of established Character in the Nation. It is a STRONG PRESUMPTION therefore against any, that they have a *strange Fire* working in them, when they are seized with SWOONINGS, and have bodily Representations of those Things, which are *spiritually* to be discerned; because these *Sights* have been common among *Enthusiasts*

The Wonderful Narrative, or, a Faithful Account of the French Prophets, their Agitations, Extasies, and Inspirations . . . (Boston, 1742), Appendix, pp. 97–104.

of all Sorts, but seldom or never among *solid* Christians. In the Beginning of the Reformation, there were Swarms of those, who pretended
to these *extraordinary* Matters; but they were always esteemed a Clog
to the Reformation, and the Disgrace of it: Nor are VISIONS and
TRANCES more common any where, than among the *Papists;* The
Lives of their Saints, as my Friend has observed, in the foregoing
Papers, are filled with Relations in this Kind.

Not that I wonder that the common People, who are unacquainted
with these Things, have been surprized; yea, even astonished at the
Shreekings, and *Faintings,* and *Agonies,* they have been Witnesses to.
And if by seeing & hearing of these Things in others, they have themselves been, in like Manner affected, it is no more than might be expected. The Frame of humane Nature is such, that it is scarce possible
it should be otherwise. And if, in many Persons, the *Imagination* has
been *heated* to a Degree sufficient to give them SIGHTS and VI-
SIONS, neither is this any more than has happened in *Thousands* of
Instances, in all *Parts* and *Ages* of the World. Nor indeed has there
appeared among us, any Thing, as to these *Extraordinaries,* but what
may be seen exemplified, yea, far exceeded in the foregoing Accounts.

But it will, doubtless, be here said, there is a great deal of that among
us, which must be acknowledged to be *true Religion,* mixt with that,
which some may think to be the *Effect* of *Imagination.* This I will not
deny: However, this is worth the remarking, that in all the before
written Instances of *Enthusiasm* and *Delusion,* there is a *mighty Shew
and Appearance, even of that, which is the Height of Religion.* Who
ever pretended to a more intimate Converse with GOD, than the
FRENCH PROPHETS? They were often in *Raptures* and *Ravishments,* from supposed divine Communications. Who ever pretended
to greater Acts of Resignation to the divine Will; of Patience under
Reproaches and Injuries; of the most tender Love, not only to one another, but to all Mankind? Who ever pretended to greater Zeal against
Sin, and for the flourishing of CHRIST'S Kingdom? Or to more in-
satiable Thirst after *continual waiting* upon GOD in the *Duties of his
Worship?* . . . And indeed, it has always been the Way of *Enthusiasm*
and *Delusion* to appear in the Guise of Religion; yea, to be so religious, as to be *superstitious,* being flamingly zealous for a being *righteous over much.*

Not that I deny, (as I hinted before) that there may be a Mixture
of *real Christianity* with great *Enthusiasm.* I am inclined to think it

was thus, at least at first, with the *Montanists* and the FRENCH PROPHETS. And I doubt not, the *unusual Appearance* among us, has been a Means to rouse many, who were before thoughtless; and to quicken many who had fallen into a Slumber. I am not against allowing, that a good Number of Sinners have (probably) been converted into Saints; and as great a Number of Saints enlivened in their Christian Work. Though I would not be thought to judge thus, from any Thing that has been observed of these Persons, either as to their *Temper* or *Behaviour,* while their Passions have been in a *violent Commotion;* much less, while they have been *seeing Visions* and *falling into Trances.* It is impossible, while in such Circumstances, but that Persons, to all Appearances, should be religious, and this to a high Degree. Thus it has been with all *Visionaries.* It was so in all the Instances related in the foregoing Papers. The only safe Way therefore, of judging in this Case, (as it appears to me) is to do it from what may be seen and known of Persons, when they have got back to a Calm and quiet State of Mind. And, if, when their Passions are subsided, and their Imaginations cooled, they now continue to discover a truly Christian *Temper* and *Conduct,* there is Reason to hope well concerning them. And this I would Hope is the Case of many among us at present.

But let not this be made a Plea to justify those Things, which are evidently the Fruit of *meer Imagination,* or something worse: For if such Things are encouraged, no one can tell what they will come to, or where they will end. We should take Warning by what we may here learn from others. And it should be a Motive to us to do so, that so many *Disorders* are already too visible among us. Such is the *praying* and *exhorting* and *singing Psalms,* at the same Time, in the same House of Worship; such is the Peoples continuing there till Midnight, and often much longer, to the breaking in upon *Family Order, Family Religion,* and *Closet Devotions;* such is the Rise of so many *weak, illiterate Lay-Preachers,* in various Parts of the Country, who, leaving their own Business, have taken upon them the Work proper to others; such is that *strange Liberty,* which some claim to themselves, of going about from *Parish* to *Parish,* calling the People off from their necessary secular Business, as tho' they were immediately commission'd of GOD, and the Work of Religion could not be carried on without their Assistance; such, in fine, is that *Spirit of Discerning,* which has put some Ministers upon such a Method of judging and censuring their *Brethren,* as has filled *several Towns* with Faction and Schism, and general Confusion. It is well known to all, who have Eyes to see, that these and such like Irregularities have appeared among us. And however light some may be disposed to make of them, they are a Reproach to *Christianity*, and evidently proceed, if not from a worse Cause, from a *Heat*, that is not *vital* and *natural*; and I shall be much mistaken, if this will not be more plain and visible, if this *Heat* rises still higher.

. . .

ANTI-ENTHUSIASTICUS

source 10

Every Idle Untruth
as a Revelation

Charles Brockwell

> Charles Brockwell was an Anglican missionary in Salem, Massachusetts, sent by the Society for the Propagation of the Gospel in Foreign Parts. His letter home to England leaves no doubt as to how he felt about enthusiasm and the Awakening.

SALEM, Feby 18, 1741/2

REVD SIR,

. . . It is impossible to relate the convulsions into which the whole Country is thrown by a set of Enthusiasts that strole about harangueing the admiring Vulgar in *extempore* nonsense, nor is it confined to these only, for Men, Women, Children, Servants, & Nigros are now become (as they phrase it) Exhorters. Their behaviour is indeed as shocking, as uncomon, their groans, cries, screams, & agonies must affect the

Extract of a letter from Charles Brockwell to the Secretary of the S. P. G., Salem, Mass., February 18, 1742, in William S. Perry, ed., *Historical Collections Relating to the American Colonial Church*, Vol. III, *Massachusetts* (Hartford, Conn.: The Church Press, 1873), pp. 453–54. Several abbreviations have been extended by the editor.

Spectators were they never so obdurate & draw tears even from the most resolute, whilst the ridiculous & frantic gestures of others cannot but excite both laughter & contempt, some leaping, some laughing, some singing, some clapping one another upon the back, &c. The tragic scene is performed by such as are entering into the pangs of the New Birth; the comic by those who are got thro' and those are so truly enthusiastic, that they tell you they saw the Joys of Heaven, can describe its situation, inhabitants, employments, & have seen their names entered into the Book of Life & can point out the writer, character & pen. And like the Papists support their fraud by recommending every dream as a Divine Vision & every idle untruth as a revelation to the admiring multitude. Their works may justly be called the works of darkness as acted in the Night & often continued to the noon of the next day & the sleep of children depriv'd of their natural rest is called a trance, & their uncouth dreams (occasion'd from the awfulness of the place, the number of Lights, the variety of action among the People, some praying, some exhorting, some swooning, &c) are deemed no less than heavenly discoveries. In Connecticut, the next Government, 'tis said many have laid their Bibles aside; and some have burnt them, as useless to those who are so plenteously fill'd with the Spirit, as to cry out Enough Lord! In short Sir, such confusion, disorder, & irregularity Eye never beheld. The illusion of the French Prophets, A[nn]o 1707, was nothing to this, & unless as to that, some unexpected accident put a period to this, I know not but this year for Enthusiasm may be as memorable as was 1692 for witchcraft for the converted cry out upon the unregenerated, as the afflicted did then upon the poor innocent wretches that unjustly suffered. Rogers of Ipswich one of this Pseudo Apostled displayed his talent in the Town on Sunday the 24th Jany & continued here so doing until the Thursday following, when he left his auditory in charge to one Elvins a Baker, who holds forth every Thursday, and tho' a fellow of consummate ignorance is nevertheless followed by great multitudes and much cried up. But I thank God that few of my Church went to hear either of them, and those that did wholly disliked them. I having taken true pains, both in publick & private to arm them against the approaching danger which was like to beset them on either side.

Provisions of all sorts have doubled the price on account of the War, & the immemorable severity of the last Winter so that it is impossible for me to subsist on my present Salary especially as Trading is now so decay'd that my People cannot punctually comply with their contract. But of this I shall give you a further account when I have the happiness of seeing you, for which I only wait the Societie's leave. I beg you would be pleased to present my duty to the Society and my Father and believe me to be,

Revd Sir,

Your most obliged Humble Servt,

CHA. BROCKWELL.

P. S. A noted Teacher in this Town is suspected of Forgery, of which if he next July Court should be found guilty, I am pretty confident many of his Congregation will draw off to the Church of England & those of the better sort.

source 11

The Great Fire of New London

Dr. Alexander Hamilton

It is unfortunate, perhaps, that opponents of the Awakening made so much of James Davenport, although one can hardly blame them, since his boorish and fanatical conduct lent itself to their charges of enthusiasm. Despite his prominence, Davenport was no match for an Edwards, a Tennent, or a Samuel Davies. Declared insane in Boston at one time and shipped home to Long Island at another, Davenport did the Awakening frequent disservice. The little episode described below, one which Davenport hoped the world would forget, occurred in New London, Connecticut, in 1743. It is related by Dr. Alexander Hamilton, a physician from Edinburgh, who toured the colonies just after the Awakening.

N: LONDON FERRY—N: LONDON

Sunday, August 26 . . . I went home att 6 o'clock, and Deacon Green's son came to see me. He entertained me with the history of the

Carl Bridenbaugh, ed., *Gentleman's Progress: The Itinerarium of Dr. Alexander Hamilton, 1744* (Chapel Hill, N.C.: University of North Carolina Press, 1948), p. 161.

behaviour of one Davenport, a fanatick preacher there who told his flock in one of his enthusiastic rhapsodies that in order to be saved they ought to burn all their idols. They began this conflagration with a pile of books in the public street, among which were Tillotson's Sermons, Beveridge's Thoughts, Drillincourt on Death, Sherlock and many other excellent authors, and sung psalms and hymns over the pile while it was a burning. They did not stop here, but the women made up a lofty pile of hoop petticoats, silk gowns, short cloaks, cambrick caps, red heeld shoes, fans, necklaces, gloves and other such aparrell, and what was merry enough, Davenport's own idol with which he topped the pile, was a pair of old, wore out, plush breaches. But this bone fire was happily prevented by one more moderate than the rest, who found means to perswade them that making such a sacrifice was not necessary for their salvation, and so every one carried of[f] their idols again, which was lucky for Davenport who, had fire been put to the pile, would have been obliged to strutt about bare-arsed, for the devil another pair of breeches had he but these same old plush ones which were going to be offered up as an expiatory sacrifise. Mr. Green took his leave of me att 10 o'clock, and I went to bed.

source 12

Be It Enacted . . .
That if Any Minister . . .
Any Person Whosoever . . .

Connecticut Assembly

> Connecticut's ministerial associations and the colony government often saw eye-to-eye about the established church in that colony. Since the religious split over the Awakening was reflected in politics, and during the upheaval the legislature was Old Light in

Charles H. Hoadly, ed., *The Public Records of the Colony of Connecticut,* VIII, *1735–1743* (Hartford, Conn.: Case, Lockwood & Brainard Co., 1874), pp. 454–57.

temperament, it was not difficult for opponents of the Revival to persuade the government to support them in their attempt to suppress it. The Assembly responded in 1742 with an act severely restricting itinerant ministers and exhorters.

AN ACT FOR REGULATING ABUSES AND CORRECTING DISORDERS IN ECCLESIASTICAL AFFAIRS

Whereas this Assembly did, by their act made in the seventh year of the reign of her late Majesty Queen Anne, establish and confirm a confession of faith, and an agreement for ecclesiastical discipline, made at Saybrook, *anno Dom.* 1708, by the reverend elders and the messengers delegated by the churches in this Colony for that purpose, under which establishment his Majesty's subjects inhabiting in this Colony have enjoyed great peace and quietness, till of late sundry persons have been guilty of disorderly and irregular practices: whereupon this Assembly, in October last, did direct to the calling of a general consociation, to sit at Guilford in November last, which said consociation was convened accordingly; at which convention it was endeavoured to prevent the growing disorders amongst the ministers that have been ordained or licenced by the associations in this government to preach, and likewise to prevent divisions and disorder among the churches and ecclesiastical societies settled by order of this Assembly: Notwithstanding which, divers of the ministers, ordained as aforesaid, and others licensed to preach by some of the associations allowed by law, have taken upon them, without any lawful call, to go into parishes immediately under the care of other ministers, and there to preach to and teach the people; and also sundry persons, some of whom are very illiterate, and have no ecclesiastical character or any authority whatsoever to preach or teach, have taken upon them publickly to teach and exhort the people in matters of religion, both as to doctrine and practice; which practices have a tendency to make divisions and contentions among the people in this Colony, and to destroy the eccelesiastical constitution established by the laws of this government, and likewise to hinder the growth and increase of vital piety and godliness in these churches, and also to introduce unqualified persons into the ministry, and more especially where one association doth intermeddle with the affairs that by the platform and agreement abovesaid, made at Saybrook aforesaid, are properly within the province and jurisdiction of another association, as to the licencing persons to preach, and ordaining ministers: Therefore,

1. *Be it enacted by the Governor, Council and Representatives, in General Court assembled, and by the authority of the same,* That if any ordained minister, or other person licenced as aforesaid to preach, shall enter into any parish not immediately under his charge, and shall there preach or exhort the people, shall be denied and secluded the benefit of any law of . . . this Colony made for the support and encouragement || of the gospel ministry, except such ordained minister or licenced person shall be expressly invited and desired so to enter into such other parish and there to preach and exhort the people, either by the settled minister and the major part of the church of said parish, or, in case there be no settled minister, then by the church or society within such parish.

· · ·

3. *And it is further enacted by the authority aforesaid,* That if any minister or ministers, contrary to the true intent and meaning of this act, shall presume to preach in any parish not under his immediate care and charge, the minister of the parish where he shall so offend, or the civil authority, or any two of the committee of such parish, shall give information thereof in writing, under their hands, to the clerk of the parish or society where such offending minister doth belong, which clerk shall receive such information, and lodge and keep the same on file in his office; and no assistant or justice of the peace in this Colony shall sign any warrant for the collecting any minister's rate, without first receiving a certificate from the clerk of the society or parish where such rate is to be collected, that no such information as is abovementioned hath been received by him or lodged in his office.

4. *And it is further enacted by the authority aforesaid,* That if any person whatsoever, that is not a settled and ordained minister, shall go into any parish and (without the express desire and invitation of the settled minister of such parish (if any there be) and the major part of the church, or if there be no such settled minister, without the express desire of the church or congregation within such parish,) publickly preach and exhort the people, shall for every such offence, upon complaint made thereof to any assistant or justice of the peace,

be bound to his peaceable and good behaviour until the next county court in that county where the offence shall be committed, by said assistant or justice of the peace, in the penal sum of one hundred pounds lawful money, that he or they will not again offend in the like kind; and the said county court may, if they see meet, further bind the person or persons offending as aforesaid to their peaceable and good behaviour during the pleasure of said court.

5. *And it is further enacted by the authority aforesaid,* That if any foreigner, or stranger that is not an inhabitant within this Colony, including as well such persons that have no ecclesiastical character or licence to preach as such as have received ordination or licence to preach by any association or presbytery, shall presume to preach, teach or publickly to exhort, in any town or society within this Colony, without the desire and licence of the settled minister and the major part of the church of such town or society, or at the call and desire of the church and inhabitants of such town or society, provided that it so happen that there is no settled minister there, that every such preacher, teacher or exhorter, shall be sent (as a vagrant person) by warrant from any one assistant or justice of the peace, from constable to constable, out of the bounds of this Colony.

source 13

The Heat and Fervour
of Their Passions

Charles Chauncy

Besides writing the best long criticism of the Awakening (*Seasonable Thoughts*), Charles Chauncy contributed the sharpest short account of the *"religious* State of Affairs" in New England. Below

[Charles Chauncy], "A Letter from a Gentleman in *Boston* to Mr. George Wishart . . . of Edinburgh . . ." Boston, August 4, 1742, published in Edinburgh in 1742; reprinted in *The Clarendon Historical Society Reprints,* Series 1, 1882–1884, No. 5 (Edinburgh, 1883), 69–84.

is his letter to George Wishart of Edinburgh which expresses with
bitter clarity his opinions respecting the Awakening and its "enthusiasts." Along with Jonathan Mayhew, Chauncy was a liberal
Congregationalist whose rational approach to religion laid the basis
for American Unitarianism, which seriously challenged orthodox
Protestantism in the late eighteenth and early nineteenth centuries.

REVEREND SIR,

I Perceive by a printed Letter from a Friend in *Edinburgh,* containing *Excerpts of Letters concerning the Success of the Gospel in
these Parts,* that marvellous Accounts have been sent Abroad of a
most glorious Work of Grace going on in *America,* as begun by Mr
Whitefield, and helpt forward by those in his way of preaching and
acting. I should be glad there had been more Truth in those Accounts.
Some of the Things related are known Falsehoods, others strangely
enlarged upon; and the Representations, in general, such, as exhibite a
wrong Idea of the *religious* State of Affairs among us. I had Thoughts
of sending you the needful Corrections of that *Pamphlet;* but my Circumstances being such, at present, as not to allow of this, must content
myself with giving you the following *summary* Narration of things as
they have appeared among us.

The Minds of People in this Part of the World, had been greatly
prepossest in Favour of Mr *Whitefield,* from the Accounts transmitted
of him, from time to time, as a *Wonder of Piety, a Man of God,* so
as *no one was like him:* Accordingly, when he came to *Town,* about
two Years since, he was received as though he had been an *Angel of
God;* yea, *a God come down in the likeness of Man.* He was strangely
flocked after by all Sorts of Persons, and much admired by the *Vulgar,*
both *great* and *small.* The *Ministers* had him in Veneration, at least
in Appearance, as much as the People; encouraged his Preaching,
attended it themselves every Day in the Week, and mostly *twice* a
Day. The grand Subject of Conversation was Mr *Whitefield,* and the
whole Business of the Town to run, from Place to Place, to hear him
preach: And, as he preach'd under such uncommon Advantages, being
high in the Opinion of the People, and having the Body of the Ministers hanging on his Lips, he soon insinuated himself still further into
the Affections of Multitudes, in so much that it became dangerous to
mention his Name, without saying something in commendation of him.

His Reception as he past through *this* and the neighbouring

Governments of *Connecticut* and *New York*, till he came to *Philadelphia*, was after much the same Manner; save only, that he met with no Admirers among the *Clergy*, unless here and there one, any where but in *Boston*: And, whether the Ministers here in general, really thought better of him than they did elsewhere, I will not be too positive to affirm. 'Tis possible, they might act as tho' they had a great Veneration for him, and so as to lead People into such an Apprehension, from *Cowardice, Affectation of Popularity*, or a *rigid Attachment to some Sentiments in Divinity*, they might imagine there was now an Advantage to establish and propagate: And I would not undertake to prove, that they might none of them be under an undue Influence from some or other of these Motives.

Much began to be now said of a *glorious Work of God* going on in the Land. *Evening-lectures* were set up in one Place and another; no less than six in this Town, *four* weekly, and *two* monthly ones, tho' the Town does not consist of above 5000 Families at the largest Computation. At some of these Lectures, it was common to mention Mr *Whitefield* by Name, both in the *Prayers* and *Sermons*; giving God Thanks for sending such an *extraordinary* Man among us, and making him the Instrument of *such extraordinary Good* to so many souls. He was indeed spoken of, as *the Angel flying through Heaven with the Everlasting Gospel*, and such Honours sacrificed to him as were due to no meer Man: Nay, to such a Height did this Spirit rise, that all who did not express a very high Thought of Mr *Whitefield*, were lookt upon with an evil Eye; and as to those who declared their Dislike of what they judged amiss of the Times, they were stigmatised as *Enemies of God and true Religion*; yea, they were openly represented, both from the *Pulpit* and the *Press*, as in danger of committing *the Sin against the Holy Ghost*, if not actually guilty even of this *unpardonable* Sin.

And here you will doubtless be disposed to enquire, what was the *great Good* this *Gentleman* was the Instrument of.

In answer whereto, I freely acknowledge, wherever he went he generally moved the *Passions*, especially of the *younger* People, and the *Females* among them; the Effect whereof was, a great Talk about Religion, together with a Disposition to be perpetually hearing Sermons, to neglect of all other Business; especially, as preach'd by those who were Sticklers for the *new Way*, as it was called. And in these things *chiefly* consisted the Goodness so much spoken of. I deny not, but there might be here and there a Person stopp'd from going on in a Course of Sin; and some might be made really better: But so far as I could judge upon the nicest Observation, the Town, in general, was not much mended in those things wherein a Reformation was greatly needed. I could not discern myself, nor many others whom I have talked with, and challenged on this Head, but that there was the same Pride and Vanity, the same Luxury and Intemperance, the same lying and tricking and cheating, as before this Gentleman came among us. There was certainly no *remarkable* Difference as to these things: And 'tis vain in any to pretend there was. This, I am sure of, there was raised such a Spirit of bitter, censorious, uncharitable judging, as was not known before; and is, wherever it reigns, a Scandal to all who call themselves Christians: Nor was it ever evident to me, but that the greatest Friends to Mr. *Whitefield* were as much puffed up with Conceit and Pride as any of their Neighbours; and as to some of them, and the more eminent too, I verily believe they possess a *worse Spirit* than before they heard of his Name, and it had been as well for them if they had never seen his Face.

But I have only entered as yet upon that Scene of Things, which has made so much Noise in the Country. A Number of Ministers in one Place and another, were by this Time formed into Mr. *Whitefield's* Temper, and began to appear and go about preaching, with a Zeal more flaming, if possible, than his. One of the most famous among these was Mr. *Gilbert Tennent,* a Man of no great Parts or Learning; his preaching was in the *extemporaneous* Way, with much Noise and little Connection. If he had taken suitable Care to prepare his Sermons, and followed Nature in the Delivery of them, he might have acquitted himself as a *middling* Preacher; but as he preached, he was an *awkward Imitator* of Mr. *Whitefield,* and too often turned off his Hearers with *mere Stuff,* which he uttered with a Spirit more bitter and uncharitable than you can easily imagine; all were *Pharisees, Hypocrites, carnal unregenerate Wretches,* both Ministers and People, who did not think just as he did, particularly as to the Doctrines of *Calvinism;* and those who opposed him, and the Work of God he was sure he was carrying on, would have opposed *Christ Jesus himself* and *his Apostles,* had they lived in their Day. This Gentleman came from *New-Brunswick* in the *Jersies* to *Boston,* in the Middle of Winter, (a Journey of more than 300 Miles) to *water the good Seed sown by Mr.* Whitefield in this Place. It was indeed at Mr. *Whitefield's* Desire, and in consequence of a Day of *Fasting and Prayer,* kept on purpose to know the Mind of God as to this Matter, that he came among us; the *Ministers in the Town,* though *fourteen* in number, being thought insufficient to carry on the *good Work* he had begun here in the Hearts of People. And though the Design this Gentleman professedly came upon, was a bare-faced Affront to the *Body of the Ministers,* yet not only the People, (which is not to be wondred at) but some of the Ministers themselves admired and followed him, as much as they had done Mr. *Whitefield* before him; and here he was, by their Encouragement, a great Part of the Winter, preaching every Day in the Week, to the taking People off from their Callings, and the introducing a Neglect of all Business but that of hearing him preach. He went from *Boston* to the *eastward,* to visit the Places where Mr. *Whitefield* had been; and on his Return home passed through the Country, preaching every where as he went along, in the same Manner, and with the same Spirit he did here in *Boston.*

And now it was, that Mr. *Whitefield's* Doctrine of *inward Feelings* began to discover itself in Multitudes, whose *sensible Perceptions* arose to such a Height, as that they *cried out, fell down, swooned away,* and, to all Appearance, were like Persons in *Fits;* and this, when the Preaching (if it may be so called) had in it as little well digested and connected good Sense, as you can well suppose. Scores in a Congregation would be in such Circumstances at a Time; nay some hundreds in some Places, to the filling the Houses of Worship with Confusion not to be expressed in Words, nor indeed conceived of by the most lively Imagination, unless where Persons have been Eye and Ear witnesses to these Things. Though I may add here, that to a Person in possession of himself, and capable of Observation, this surprising Scene of Things may be accounted for: The *Speaker* delivers himself, with the *greatest Vehemence* both of *Voice* and *Gesture,* and in the most *frightful Language* his Genius will allow of. If this has its intended Effect upon *one* or *two weak Women,* the Shrieks catch from one to another, till a great Part of the Congregation is affected; and some are in the Thought, that it may be too common for those *zealous in the new Way to cry out themselves,* on purpose to move others, and bring forward a *general Scream. Visions* now became common, and *Trances* also, the Subjects of which were in their own Conceit transported from Earth to Heaven, where they saw and heard most glorious Things; conversed with *Christ* and *holy Angels;* had opened to them the *Book of Life,* and were permitted to read the names of persons there, and the like. And what is a singular Instance (so far as I remember) of the working of Enthusiasm, *laughing, loud hearty laughing,* was one of the Ways in which our *new Converts,* almost every where, were wont to join together in expressing their Joy at the Conversion of others.

'Tis scarce imaginable what Excesses and Extravagancies People were running into, and even encouraged in; being told such Things were Arguments of the *extraordinary Presence of the Holy Ghost* with them. The same Houses of Worship were scarce emptied Night nor Day for a Week together, and unheard of Instances of supposed Religion were carried on in them, some would be *praying,* some *exhorting,* some *singing,* some *clapping their Hands,* some *laughing,* some *crying,* some *shrieking and roaring out;* and so invincibly set were they in these Ways, especially when encouraged by any Ministers, (as was too often the Case) that it was a vain Thing to argue with them, to shew them the Indecency of such Behaviour; and whoever indeed made an Attempt this Way, might be sure aforehand of being called an *Opposer* of the *Spirit,* and a *Child of the Devil.*

At these Times there were among the People what we call here EXHORTERS; these are such as are esteemed to be *Converts* in the *new Way.* Sometimes they are *Children, Boys* and *Girls,* sometimes *Women;* but most commonly *raw, illiterate, weak* and *conceited young Men,* or *Lads.* They pray with the People, call upon them to come to Christ, tell them they are dropping into Hell, and take upon them what they imagine is the Business of preaching. They are generally much better thought of than any Ministers, except those in the *new Way,* I mean by the Friends to the *Extraordinaries* prevalent in the Land; and they are the greatest promoters of them. 'Tis indeed at the *Exhortations* of these poor ignorant Creatures, that there is ordinarily the most Noise and Confusion: And what may be worth a particular Remark, 'tis *seldom* there are any great Effects wrought, till the Gloominess of the Night comes on. It is in the *Evening,* or more late in the *Night,* with only a *few Candles* in a *Meeting-house,* that there is the *screaming* and *shrieking* to the greatest Degree; and the Persons thus affected are generally *Children, young People,* and *Women.* Other Instances there may have been, but they are more rare; these bear the chief Part.

I shall here insert a Paragraph of a Letter sent me by a Friend living at *Newhaven,* the seat of one of our *Colleges,* a Gentleman of known Integrity and Veracity, giving an Account of the Managements of one of the Preachers of Mr. *Whitefield's* making, with the Appearance following thereupon. Says he, "After the Conclusion of the Exercises usual in our religious Assemblies, he came down from the *Pulpit* into the *Deacon's* Seat. His Exercises were, 1. *Short Prayers;* wherein he used very uncommon Expressions, and such as had no Tendency, at least in my Mind, to excite Devotion; which he delivered with a boisterous Voice, and in a Manner to me very disagreeable. 2. *Singing Psalms* and *Hymns;* which he him self repeated with an awful Tone and frightful Gestures. 3. *Exhorting,* as they called it: to which many *Laymen* were admitted as *Assistants.* In performing these Exercises they observed no stated Method, but proceeded as their present Thought or Fancy led them: And by this means the Meetinghouse would be filled with what I could not but judge great Confusion and Disorder; for the whole House would many times seem to be in a perfect Hubbub, and People filled with Consternation. These Meetings they would continue till 10, 11, 12 o'Clock at Night; in the midst of them sometimes 10, 20, 30, and sometimes many more would *scream* and *cry out,* or send forth the most *lamentable Groans,* whilst others made great Manifestations of Joy, by *clapping their Hands,* uttering *extatick Expressions, singing Psalms,* and *inviting* and *exhorting* others. Some would *swoon away* under the Influence of distressing Fears, and others *swallowed up with insupportable Joy.* While some were *fainting,* others laboured under *convulsive Twitches of Body,* which they said were involuntary. But in vain shall I pretend to describe all the Proceedings at those Meetings. But what appeared to me most dangerous and hurtful was, that very much Stress was laid on these *Extraordinaries,* as tho' they were *sure Marks,* or, at least *sufficient Evidences* of a just Conviction of Sin on the one Hand; or, on the other, of that Joy which there is in believing, and so of an Interest in the Favour of God."

You may be ready perhaps to think I have here given you a romantick Representation of Things; but it is the real Truth of the

Case without a Figure; yea, this has been the Appearance in all Parts of the Land more or less, and so known to have been so, that there is no room for Debate upon the Matter: Nay, those who are Friends to the *new Way* were *once* so far from being ashamed of these Things, that they boasted of them, and entertained an ill Opinion of all who did not speak of them as *Evidences* of the *wonderful Power of the Spirit of God*: I say, they *at first* boasted of these Things, and some of them do so still; though the Generality have begun, for some time, to speak publickly of the *Subtility of Satan,* to tell People he may appear as *an Angel of Light,* and to warn them against being carried away by his Devices. Nay Mr. *Tennent* himself, one of the main Instruments of all our Disorders, has, in a couple of Letters to some of his Friends, published in the *Prints* (*a*), expressed his Fears lest the Churches should be undone with a *Spirit of Enthusiasm,* and *these Exhorters* which have risen up everywhere in the Land. He seems indeed to have quite turned about: The Reason whereof may be this; the *Moravians* who came to *Philadelphia* with Count *Zinzendorf,* have been among his People, and managed with them as he did elsewhere, and brought the like Confusion among them; and now he cries out of Danger, and expresses himself much as those did, whom before he had sent to the Devil by wholesale.

Various are the Sentiments of Persons about this *unusual Appearance* among us. Some think it to be a *most wonderful Work of God's Grace;* others a *most wonderful Spirit of Enthusiasm;* some think there is a *great deal of Religion,* with some *small Mixture* of Extravagance; others a *great deal of Extravagance* with some *small Mixture* of that which may be called *good;* some think the *Country* was never in such a *happy* State on a *religious* account, others that it was never in a *worse.*

For my self, I am among those who are clearly in the Opinion, that there never was such a *Spirit* of *Superstition* and *Enthusiasm* reigning in the Land before; never such *gross Disorders* and *barefaced Affronts* to *common Decency;* never such *scandalous Reproaches* on the *Blessed Spirit,* making him the Author of the greatest *Irregularities* and *Confusions:* Yet, I am of Opinion also, that the Appearances among us (so much out of the ordinary Way, and so unaccountable to persons not acquainted with the History of the World) have been the Means of awakening the Attention of many; and a good Number,

I hope, have settled into a truly *Christian* Temper: Tho' I must add, at the same time, that I am far from thinking, that the Appearance, in *general*, is any other than the Effect of *enthusiastick Heat*. The Goodness that has been so much talked of, 'tis plain to me, is nothing more, in general, than a *Commotion in the Passions*. I can't see that Men have been made *better*, if hereby be meant, their being formed to a nearer Resemblance to the *Divine Being* in *moral Holiness*. 'Tis not evident to me, that Persons, generally, have a better Understanding of Religion, a better Government of their Passions, a more Christian Love to their Neighbour, or that they are more decent and regular in their Devotions towards God. I am clearly of the Mind, they are worse in all these Regards. They place their Religion so much in the *Heat* and *Fervour* of their *Passions*, that they too much neglect their *Reason* and *Judgment*: And instead of being more kind and gentle, more full of Mercy and good Fruits, they are more bitter, fierce and implacable. And what is a *grand discriminating Mark of this Work*, where-ever it takes place, is, that it makes Men *spiritually proud* and *conceited* beyond Measure, infinitely *censorious* and *uncharitable*, to *Neighbours*, to *Relations*, even the nearest and dearest; to *Ministers* in an especial Manner; yea, to all Mankind, who are not as they are, and don't think and act as they do: And there are few places where *this Work* has been in any *remarkable* manner, but they have been filled with Faction and Contention; yea, in some, they have divided into Parties, and openly and scandalously separated from one another.

Truly the Accounts sent Abroad, were sent too soon; too soon, I am satisfied, to reflect Honour upon the Persons who wrote them: And they betray such a want of Judgment, as I was really sorry to see them falling into. There are few Persons now, perhaps none but such as are evidently over-heated, but begin to see that Things have been carried too far, and that the Hazard is great, unless God mercifully interpose, lest we should be over-run with *Enthusiasm*. And to speak the plain Truth, my Fear is, lest the End of these things should be *Quakerism* and *Infidelity*: These we have now chiefly to guard against.

A particular Account of one Mr *James Davenport*, with his *strange Conduct* in *Town* and *elsewhere*, I doubt not would have been agreeable: But I have exceeded already. He is the *wildest Enthusiast* I ever saw, and acts in the wildest manner; and yet, he is vindicated by some in all his Extravagancies.

I now beg Pardon, Sir, for thus trespassing upon your Patience. As Mr. *Whitefield* has been in *Scotland,* and *human Nature* is the *same every where;* this Narration of the Effects he has been the Instrument of producing here, may excite your Zeal to guard the People in time against any such Extravagancies, if there should be Danger of them where you may be concerned. I am,

Reverend Sir,
With all due Regard, &c.

Boston, August 4.
1742.

POSTSCRIPT

One thing I forgot to mention, which yet is worthy of Note. It is, That in many Places, where Persons can't any longer *scream out* for themselves, supposing themselves to have got into a *converted State,* they will *scream* and *cry out,* and make as great a Noise as they can on the account of others, for their *unconverted Neighbours,* that are in a State of Sin, and going to Hell. This begins now to be common; it has been practised in many Places.

FINIS

part three

THERE IS INDEED
SUCH A THING
AS TRUE
EXPERIMENTAL RELIGION

source 14

A Thousand Imprudences
Will Not Prove a Work
to Be Not of the Spirit
of God

Jonathan Edwards

Jonathan Edwards was on the spot when he delivered this sermon at Yale College in 1741. As tension between camps mounted over the issues of the Awakening, most New England ministers were eager to learn which side of the controversy the formidable Edwards would support, particularly with regard to the issue of enthusiasm. He not only satisfied both friends and opponents as to where he stood, but he gave all within earshot a brilliant lecture on "vital religion."

SECTION I

Negative Signs; or, What are no signs by which
we are to judge of a work—and especially,
What are no evidences that a work is not
from the Spirit of God.

I. Nothing can be certainly concluded from this, That a work is carried on in a way very unusual and extraordinary; provided the variety or difference be such, as may still be comprehended within the limits of Scripture rules. What the church has been used to, is not a rule by which we are to judge; because there may be new and extraordinary works of God, and he has heretofore evidently wrought in

Jonathan Edwards, "The Distinguishing Marks of a Work of the Spirit of God," *The Works of President Edwards,* I (New York: Leavitt and Company, 1851), pp. 526–38, 559–62.

an extraordinary manner. He has brought to pass new things, strange works; and has wrought in such a manner as to surprise both men and angels. And as God has done thus in times past, so we have no reason to think but that he will do so still. The prophecies of Scripture give us reason to think that God has things to accomplish, which have never yet been seen. No deviation from what has hitherto been usual, let it be never so great, is an argument that a work is not from the Spirit of God, if it be no deviation from his prescribed rule. The Holy Spirit is sovereign in his operation; and we know that he uses a great variety; and we cannot tell how great a variety he may use, within the compass of the rules he himself has fixed. We ought not to limit God where he has not limited himself.

Therefore it is not reasonable to determine that a work is not from God's Holy Spirit because of the extraordinary degree in which the minds of persons are influenced. If they seem to have an extraordinary conviction of the dreadful nature of sin, and a very uncommon sense of the misery of a Christless condition—or extraordinary views of the certainty and glory of divine things,—and are proportionably moved with very extraordinary affections of fear and sorrow, desire, love, or joy: or if the apparent change be very sudden, and the work be carried on with very unusual swiftness—and the persons affected are very numerous, and many of them are very young, with other unusual circumstances, not infringing upon Scripture marks of a work of the Spirit—these things are no argument that the work is not of the Spirit of God.

· · ·

There is a great aptness in persons to doubt of things that are strange; especially elderly persons, to think that to be right which they have never been used to in their day, and have not heard of in the days of their fathers. But if it be a good argument that a work is not from the Spirit of God, that it is very unusual, then it was so in the apostles' days. The work of the Spirit then, was carried on in a manner that, in very many respects, was altogether new; such as never had been seen or heard since the world stood. The work was then carried on with more visible and remarkable power than ever; nor had there been seen before such mighty and wonderful effects of the Spirit of God in sudden

changes, and such great engagedness and zeal in great multitudes—
such a sudden alteration in towns, cities, and countries; such a swift
progress, and vast extent of the work—and many other extraordinary
circumstances might be mentioned. The great unusualness of the work
surprised the Jews; they knew not what to make of it, but could not
believe it to be the work of God: many looked upon the persons that
were the subjects of it as bereft of reason; as you may see in Acts ii. 13,
xxvi. 24, and 1 Cor. iv. 10.

And we have reason from Scripture prophecy to suppose, that at
the commencement of that last and greatest outpouring of the Spirit
of God, that is to be in the latter ages of the world, the manner of the
work will be very extraordinary, and such as never has yet been seen;
so that there shall be occasion then to say, as in Isa. lxvi. 8, "Who hath
heard such a thing? Who hath seen such things? Shall the earth be
made to bring forth in one day? Shall a nation be born at once? for as
soon as Zion travailed, she brought forth her children." It may be rea-
sonably expected that the extraordinary manner of the work then, will
bear some proportion to the very extraordinary events, and that glorious change in the state of the world, which God will bring to pass
by it.

II. A work is not to be judged of by any effects on the bodies of
men; such as tears, trembling, groans, loud outcries, agonies of body,
or the failing of bodily strength. The influence persons are under, is
not to be judged of one way or other, by such effects on the body; and
the reason is, because the Scripture nowhere gives us any such rule.
We cannot conclude that persons are under the influence of the true
Spirit because we see such effects upon their bodies, because this is not
given as a mark of the true Spirit; nor on the other hand, have we any
reason to conclude, from any such outward appearances, that persons
are not under the influence of the Spirit of God, because there is no rule
of Scripture given us to judge of spirits by, that does either expressly
or indirectly exclude such effects on the body, nor does reason exclude
them. It is easily accounted for from the consideration of the nature
of divine and eternal things, and the nature of man, and the laws
of the union between soul and body, how a right influence, a true
and proper sense of things, should have such effects on the body,
even those that are of the most extraordinary kind, such as taking
away the bodily strength, or throwing the body into great agonies,

and extorting loud outcries. There are none of us but do suppose, and would have been ready at any time to say it, that the misery of hell is doubtless so dreadful, and eternity so vast, that if a person should have a clear apprehension of that misery as it is, it would be more than his feeble frame could bear, and especially if at the same time he saw himself in great danger of it, and to be utterly uncertain whether he should be delivered from it, yea, and to have no security from it one day or hour. If we consider human nature, we must not wonder, that when persons have a great sense of that which is so amazingly dreadful, and also have a great view of their own wickedness and God's anger, that things seem to them to forebode speedy and immediate destruction. We see the nature of man to be such that when he is in danger of some terrible calamity to which he is greatly exposed, he is ready upon every occasion to think, that *now* it is coming.—When persons' hearts are full of fear, in time of war, they are ready to tremble at the shaking of a leaf, and to expect the enemy every minute, and to say within themselves, *now* I shall be slain. If we should suppose that a person saw himself hanging over a great pit, full of fierce and glowing flames, by a thread that he knew to be very weak, and not sufficient to bear his weight, and knew that multitudes had been in such circumstances before, and that most of them had fallen and perished, and saw nothing within reach, that he could take hold of to save him, what distress would he be in! How ready to think that *now* the thread was breaking, that now, *this minute,* he should be swallowed up in those dreadful flames! And would not he be ready to cry out in such circumstances? How much more those that see themselves in this manner hanging over an infinitely more dreadful pit, or held over it in the hand of God, who at the same time they see to be exceedingly provoked! No wonder that the wrath of God, when manifested but a little to the soul, overbears human strength. . . .

Some object against such extraordinary appearances, that we have no instances of them recorded in the New Testament, under the extraordinary effusions of the Spirit. Were this allowed, I can see no force in the objection, if neither reason, nor any rule of Scripture, exclude such things; especially considering what was observed under the foregoing particular. I do not know that we have any express mention in the New Testament of any person's weeping, or groaning, or sighing through fear of hell, or a sense of God's anger; but is there any body so foolish as from hence to argue, that in whomsoever these things appear, their convictions are not from the Spirit of God? And the reason why we do not argue thus, is, because these are easily accounted for, from what we know of the nature of man, and from what the Scripture informs us in general, concerning the nature of eternal things, and the nature of the convictions of God's Spirit; so that there is no need that any thing should be said in particular concerning these external, circumstantial effects. Nobody supposes that there is any need of express scripture for every external, accidental manifestation of the inward motion of the mind: and though such circumstances are not particularly recorded in sacred history, yet there is a great deal of reason to think, from the general accounts we have, that it could not be otherwise than that such things must be in those days. And there is also reason to think, that such great outpouring of the Spirit was not wholly without those more extraordinary effects on persons' bodies. . . .—We read of the disciples, Matt. xiv. 26, that when they saw Christ coming to them in the storm, and took him for some terrible enemy, threatening their destruction in that storm, "they cried out for fear." Why then should it be thought strange, that persons should cry out for fear, when God appears to them, as a terrible enemy, and they see themselves in great danger of being swallowed up in the bottomless gulf of eternal misery? The spouse, once and again, speaks of herself as overpowered with the love of Christ, so as to weaken her body, and make her faint. Cant. ii. 5, "Stay me with flagons, comfort me with apples; for I am sick of love." And chap. v. 8, "I charge you, O ye daughters of Jerusalem, if ye find my Beloved, that ye tell him that I am sick of love." From whence we may at least argue, that such an effect may well be supposed to arise from such a cause in the saints, in some cases, and that such an effect will sometimes be seen in the church of Christ.

It is a weak objection, that the impressions of enthusiasts have a great effect on their bodies. That the Quakers used to tremble, is no argument that Saul, afterwards Paul, and the jailer, did not tremble from real convictions of conscience. Indeed all such objections from effects on the body, let them be greater or less, seem to be exceeding frivolous; they who argue thence, proceed in the dark, they know not what ground they go upon, nor by what rule they judge. The root and course of things is to be looked at, and the nature of the operations and affections are to be inquired into, and examined by the rule of God's word, and not the motions of the blood and animal spirits.

III. It is no argument that an operation on the minds of people is not the work of the Spirit of God, that it occasions a great deal of noise about religion. For though true religion be of a contrary nature to that of the Pharisees—which was ostentatious, and delighted to set itself forth to the view of men for their applause—yet such is human nature, that it is morally impossible there should be a great concern, strong affection, and a general engagedness of mind amongst a people without causing a notable, visible, and open commotion and alteration amongst that people.—Surely, it is no argument that the minds of persons are not under the influence of God's Spirit, that they are very much moved: for indeed spiritual and eternal things are so great, and of such infinite concern, that there is a great absurdity in men's being but moderately moved and affected by them; and surely it is no argument that they are not moved by the Spirit of God, that they are affected with these things in some measure as they deserve, or in some proportion to their importance. And when was there ever any such thing since the world stood, as a people in general being greatly affected in any affair whatsoever, without noise or stir? The nature of man will not allow it.

·   ·   ·

IV. It is no argument that an operation on the minds of a people, is not the work of the Spirit of God, that many who are the subjects of it, have great impressions made on their imaginations. That persons have many impressions on their imaginations, does not prove that they have nothing else. It is easy to be accounted for, that there should be much of this nature amongst a people, where a great multitude of all kinds of constitutions have their minds engaged with intense thought and strong affections about invisible things; yea, it would be strange if there should not. Such is our nature, that we cannot think of things invisible, without a degree of imagination. I dare appeal to any man, of the greatest powers of mind, whether he is able to fix his thoughts on God, or Christ, or the things of another world, without imaginary ideas attending his meditations? And the more engaged the mind is, and the more intense the contemplation and affection, still the more lively and strong the imaginary idea will ordinarily be; especially when attended with surprise. And this is the case when the mental prospect is very new, and takes strong hold of the passions, as fear or joy; and when the change of the state and views of the mind is sudden, from a contrary extreme, as from that which was extremely dreadful, to that which is extremely ravishing and delightful. And it is no wonder that many persons do not well distinguish between that which is imaginary and that which is intellectual and spiritual; and that they are apt to lay too much weight on the imaginary part, and are most ready to speak of that in the account they give of their experiences, especially persons of less understanding and of distinguishing capacity.

As God has given us such a faculty as the imagination, and so made us that we cannot think of things spiritual and invisible, without some exercise of this faculty; so, it appears to me, that such is our state and nature, that this faculty is really subservient and helpful to the other faculties of the mind, when a proper use is made of it; though oftentimes, when the imagination is too strong, and the other faculties weak, it overbears, and disturbs them in their exercise. It appears to me manifest, in many instances with which I have been acquainted, that God has really made use of this faculty to truly divine purposes; especially in some that are more ignorant. God seems to condescend to their circumstances, and deal with them as babes; as of old he instructed his church, whilst in a state of ignorance and minority, by types and outward representations. I can see nothing unreasonable in such a position. Let others who have much occasion to deal with souls in spiritual concerns, judge whether experience does not confirm it.

It is no argument that a work is not of the Spirit of God, that some who are the subjects of it have been in a kind of ecstasy, wherein they have been carried beyond themselves, and have had their minds transported into a train of strong and pleasing imaginations, and a kind of visions, as though they were rapt up even to heaven, and there saw glorious sights. I have been acquainted with some such instances, and I see no need of bringing in the help of the devil into the account that we give of these things, nor yet of supposing them to be of the same nature with the visions of the prophets, or St. Paul's rapture into paradise. Human nature, under these intense exercises and affections, is all that need be brought into the account. If it may be well accounted for, that persons under a true sense of a glorious and wonderful greatness and excellency of divine things, and soul-ravishing views of the beauty and love of Christ, should have the strength of nature overpowered, as I have already shown that it may; then I think it is not at all strange, that amongst great numbers that are thus affected and overborne, there should be some persons of particular constitutions that should have their imaginations thus affected. The effect is no other than what bears a proportion and analogy to other effects of the strong exercise of their minds. It is no wonder, when the thoughts are so fixed, and the affections so strong—and the whole soul so engaged, ravished, and swallowed up—that all other parts of the body are so affected, as to be deprived of their strength, and the whole frame ready to dissolve. Is it any wonder that, in such a case, the brain in particular (especially in some constitutions), which we know is most especially affected by intense contemplations and exercises of mind, should be so affected, that its strength and spirits should for a season be diverted, and taken off from impressions made on the organs of external sense, and be wholly employed in a train of pleasing delightful imaginations, corresponding with the present frame of the mind? Some are ready to interpret such things wrong, and to lay too much weight on them, as prophetical visions, divine revelations, and sometimes significations from heaven of what shall come to pass; which the issue, in some instances I have known, has shown to be otherwise. But yet, it appears to me that such things are evidently sometimes from the Spirit of God, though indirectly; that is, their extraordinary frame of mind, and that strong and lively sense of divine things which is the occasion of them, is from his Spirit; and also as the mind continues in its holy frame, and retains a divine sense of the excellency of spiritual things even in its rapture; which holy frame and sense is from the Spirit of God, though the imaginations that attend it are but accidental, and therefore there is commonly something or other in them that is confused, improper, and false.

V. It is no sign that a work is not from the Spirit of God, that example is a great means of it. It is surely no argument that an effect is not from God, that means are used in producing it; for we know that it is God's manner to make use of means in carrying on his work in the world, and it is no more an argument against the divinity of an effect, that this means is made use of, than if it was by any other means. It is agreeable to Scripture that persons should be influenced by one another's good example. . . .

And as it is a *Scriptural* way of carrying on God's work, by example, so it is a *reasonable* way. It is no argument that men are not influenced by reason, that they are influenced by example. This way of persons holding forth truth to one another, has a tendency to enlighten the mind, and to convince reason. None will deny but that for persons to signify things one to another by words, may rationally be supposed to tend to enlighten each other's minds. But the same thing may be signified by actions, and signified much more fully and effectually. Words are of no use any otherwise than as they convey our own ideas to others; but actions, in some cases, may do it much more fully. There is a language in actions; and in some cases, much more clear and convincing than in words. It is therefore no argument against the goodness of the effect, that persons are greatly affected by seeing others so; yea, though the impression be made only by seeing the tokens of great and extraordinary affection in others in their behavior, taking for granted what they are affected with, without hearing them say one word. There may be language sufficient in such a case in their behavior only, to convey their minds to others, and to signify to them their sense of things more than can possibly be done by words only. If a person should see another under extreme bodily torment, he might receive much clearer ideas, and more convincing evidence of what he suffered by his actions in his misery, than he could do only by the words of an unaffected indifferent relater. In like manner he might receive a greater idea of any thing that is excellent and very delightful, from the behavior of one that is in actual enjoyment, than by the dull narration of one which is inexperienced and insensible himself. I desire that this matter may be examined by the strictest reason.—Is it not manifest, that effects produced in persons' minds are rational, since not only weak and ignorant people are much influenced by example, but also those that make the greatest boast of strength of reason, are more influenced by reason held forth in this way, than almost any other way. Indeed the religious affections of many when raised by this means, as by hearing the word preached, or any other means, may prove flashy, and soon vanish, as Christ represents the stony-ground hearers; but the affections of some thus moved by example, are abiding, and prove to be of saving issue.

.   .   .

It is no valid objection against examples being so much used, that the Scripture speaks of the word as the principal means of carrying on God's work; for the word of God is the principal means, nevertheless, by which other means operate and are made effectual. Even the sacraments have no effect but by the word; and so it is that example becomes effectual; for all that is visible to the eye is unintelligible and vain, without the word of God to instruct and guide the mind. It is the word of God that is indeed held forth and applied by example, as the word of the Lord sounded forth to other towns in Macedonia, and Achaia, by the example of those that believe in Thessalonica.

. • •

VI. It is no sign that a work is not from the Spirit of God, that many, who seem to be the subjects of it, are guilty of great imprudences and irregularities in their conduct. We are to consider that the end for which God pours out his Spirit, is to make men holy, and not to make them politicians. It is no wonder that, in a mixed multitude of all sorts— wise and unwise, young and old, of weak and strong natural abilities, under strong impressions of mind—there are many who behave themselves imprudently. There are but few that know how to conduct themselves under vehement affections of any kind, whether of a temporal or spiritual nature; to do so requires a great deal of discretion, strength, and steadiness of mind. A thousand imprudences will not prove a work to be not of the Spirit of God; yea, if there be not only imprudences, but many things prevailing that are irregular, and really contrary to the rules of God's holy word. That it should be thus may be well accounted for from the exceeding weakness of human nature, together with the remaining darkness and corruption of those that are yet the subjects of the saving influences of God's Spirit, and have a real zeal for God.

We have a remarkable instance, in the New Testament, of a people that partook largely of that great effusion of the Spirit in the apostles' days, among whom there nevertheless abounded imprudences and great irregularities; viz., the church at Corinth. There is scarcely any church more celebrated in the New Testament for being blessed with large measures of the Spirit of God, both in his ordinary influences, in convincing and converting sinners, and also in his extraordinary and miraculous gifts; yet what manifold imprudences, great and sinful irregularities, and strange confusion did they run into, at the Lord's supper, and in the exercise of church discipline! To which may be added, their indecent manner of attending other parts of public worship, their jarring and contention about their teachers, and even the exercise of their extraordinary gifts of prophecy, speaking with tongues, and the like, wherein they spake and acted by the immediate inspiration of the Spirit of God.

And in particular, it is no evidence that a work is not of God, if many who are either the subjects or the instruments of it, are guilty of too great forwardness to censure others as unconverted. For this may be through mistakes they have embraced concerning the marks by which they are to judge of the hypocrisy and carnality of others; or from not duly apprehending the latitude the Spirit of God uses in the methods of his operations; or, from want of making due allowance for that infirmity and corruption that may be left in the hearts of the saints; as well as through want of a due sense of their own blindness and weakness, and remaining corruption, whereby spiritual pride may have a secret vent this way, under some disguise, and not be discovered. If we allow that truly pious men may have a great deal of remaining blindness and corruption, and may be liable to mistakes about the marks of hypocrisy, as undoubtedly all will allow, then it is not unaccountable that they should sometimes run into such errors as these. It is as easy, and upon some accounts more easy to be accounted for, why the remaining corruption of good men should sometimes have an unobserved vent this way than most other ways; and without doubt (however lamentable) many holy men have erred in this way.

Lukewarmness in religion is abominable, and zeal an excellent grace, yet above all other Christian virtues, this needs to be strictly watched and searched; for it is that with which corruption, and particularly pride and human passion, is exceedingly apt to mix unobserved. And it is observable, that there never was a time of great reformation, to cause a revival of zeal in the church of God, but that it has been attended, in some notable instances, with irregularity, and a running out some way or other into an undue severity. . . . So in the church of Corinth, they had got into a way of extolling some ministers, and censuring others, and were puffed up one against another; but yet these things were no sign that the work then so wonderfully carried on, was not the work of God. And after this, when religion was still greatly flourishing in the world, and a Spirit of eminent holiness and zeal prevailed in the Christian church, the zeal of Christians ran out into a very improper and undue severity, in the exercise of church discipline towards delinquents. In some cases they would by no means admit them into their charity and communion though they appeared never so humble and penitent. And in the days of Constantine the Great, the zeal of Christians against heathenism ran out into a degree of persecution. So in that glorious revival of religion, at the reformation, zeal in many instances appeared in a very improper severity, and even a degree of persecution; yea, in some of the most eminent reformers; as in the great Calvin in particular. And many in those days of the flourishing of vital religion, were guilty of severely censuring others that differed from them in opinion in some points of divinity.

VII. Nor are many errors in judgment, and some delusions of Satan intermixed with the work, any argument that the work in general is not of the Spirit of God. However great a spiritual influence may be, it is not to be expected that the Spirit of God should be given now in the same manner as to the apostles, infallibly to guide them in points of Christian doctrine, so that what they taught might be relied on as a rule to the Christian church. And if many delusions of Satan appear, at the same time that a great religious concern prevails, it is not an argument that the work in general is not the work of God. . . . Yea, the same persons may be the subjects of much of the influences of the Spirit of God, and yet in some things be led away by the delusions of Satan, and this be no more of paradox than many other things that are true of real saints, in the present state, where grace dwells with so much corruption, and the new man and the old man subsist together in the same person; and the kingdom of God and the kingdom of the devil remain for a while together in the same heart. Many godly persons have undoubtedly in this and other ages, exposed themselves to woful delusions, by an aptness to lay too much weight on impulses and impressions, as if they were immediate revelations from God, to signify something future, or to direct them where to go, and what to do.

VIII. If some, who were thought to be wrought upon, fall away into gross errors, or scandalous practices, it is no argument that the work in general is not the work of the Spirit of God. That there are some counterfeits, is no argument that nothing is true: such things are always expected in a time of reformation. If we look into church history, we shall find no instance of any great revival of religion, but what has been attended with many such things. Instances of this nature in the apostles' days were innumerable; some fell away into gross heresies, others into vile practices, though they seemed to be the subjects of a work of the Spirit—and were accepted for a while amongst those that were truly so, as their brethren and companions—and were not suspected till they went out from them. And some of these were teachers and officers— and eminent persons in the Christian church—whom God had endowed with miraculous gifts of the Holy Ghost; as appears by the beginning of the 6th chapter of the Hebrews. . . . So in the time of the reformation from popery, how great was the number of those who for a while seemed to join with the reformers, yet fell away into the grossest and most absurd errors, and abominable practices. And it is particularly observable, that in times of great pouring out of the Spirit to revive religion in the world, a number of those who for a while seemed to partake in it, have fallen off into whimsical and extravagant errors, and gross enthusiasm, boasting of high degrees of spirituality and perfection, censuring and condemning others as carnal. Thus it was with the Gnostics in the apostles' times; and thus it was with several sects at the Reformation, as Anthony Burgess observes in his book called Spiritual Refinings, Part I. Serm. 23. p. 132: "The first worthy reformers, and glorious instruments of God, found a bitter conflict herein, so that they were exercised not only with formalists, and traditionary papists on the one side, but men that pretended themselves to be more enlightened than the reformers were, on the other side: hence they called those that did adhere to the Scripture, and would try revelations by it, Literists and Vowelists, as men acquainted with the words and vowels of the Scripture, having nothing of the Spirit of God: and wheresoever in any town, the true doctrine of the gospel brake forth to the displacing of popery, presently such opinions arose like tares that came up among the good wheat; whereby great divisions were raised, and the reformation made abominable and odious to the world; as if that had been the sun to give heat and warmth to those worms and serpents to crawl out of the ground. Hence they inveighed against Luther, and said he had only promulgated a carnal gospel."—Some of the leaders of those wild enthusiasts had been for a while highly esteemed by the first reformers, and peculiarly dear to them.—Thus also in England, at the time when vital religion much prevailed in the days of King Charles I. the interregnum, and Oliver Cromwell, such things as these abounded. And so in New England, in her purest days, when vital piety flourished, such kind of things as these broke out. Therefore the devil's sowing of such tares is no proof that a true work of the Spirit of God is not gloriously carried on.

IX. It is no argument that a work is not from the Spirit of God, that it seems to be promoted by ministers insisting very much on the terrors of God's holy law, and that with a great deal of pathos and earnestness. If there be really a hell of such dreadful and never-ending torments, as is generally supposed, of which multitudes are in great danger—and into which the greater part of men in Christian countries do actually from generation to generation fall, for want of a sense of its terribleness, and so for want of taking due care to avoid it—then why is it not proper for those who have the care of souls to take great pains to make men sensible of it? Why should they not be told as much of the truth as can be? If I am in danger of going to hell, I should be glad to know as much as possibly I can of the dreadfulness of it. If I am very prone to neglect due care to avoid it, he does me the best kindness, who does most to represent to me the truth of the case, that sets forth my misery and danger in the liveliest manner.

·   ·   ·

When ministers preach of hell, and warn sinners to avoid it, in a cold manner—though they may say in words that it is infinitely terrible—they contradict themselves. For actions, as I observed before, have a language as well as words. If a preacher's words represent the sinner's state as infinitely dreadful, while his behavior and manner of speaking contradict it—showing that the preacher does not think so—he defeats his own purpose; for the language of his actions, in such a case, is much more effectual than the bare signification of his words. Not that I think that the law only should be preached: ministers may preach other things too little. The gospel is to be preached as well as the law, and the law is to be preached only to make way for the gospel, and in order that it may be preached more effectually. The main work of ministers is to preach the gospel: "Christ is the end of the law for righteousness." So that a minister would miss it very much if he should insist so much on the terrors of the law, as to forget his Lord, and neglect to preach the gospel; but yet the law is very much to be insisted on, and the preaching of the gospel is like to be in vain without it.

SECTION II

. . .

Moreover, seeing inspiration is not to be expected, *let us not despise human learning.* They who assert that human learning is of little or no use in the work of the ministry, do not well consider what they say; if they did, they would not say it. By human learning I mean, and suppose others mean, the improvement of common knowledge by human and outward means. And therefore to say, that human learning is of no use, is as much as to say that the education of a child, or that the common knowledge which a grown man has more than a little child, is of no use. At this rate, a child of four years old is as fit for a teacher in the church of God, with the same degree of grace—and capable of doing as much to advance the kingdom of Christ, by his instruction—as a very knowing man of thirty years of age. If adult persons have greater ability and advantage to do service, because they have more knowledge than a little child, then doubtless if they have more human knowledge still, with the same degree of grace, they would have still greater ability and advantage to do service. An increase of knowledge, without doubt, increases a man's advantage either to do good or hurt, according as he is disposed. It is too manifest to be denied, that God made great use of human learning in the apostle Paul, as he also did in Moses and Solomon.

And if knowledge, obtained by human means, is not to be despised, then it will follow that the means of obtaining it are not to be neglected, viz., *study;* and that this is of great use in order to a preparation for publicly instructing others. And, though having the heart full of the powerful influences of the Spirit of God, may at some time enable persons to speak profitably, yea, very excellently, without study; yet this will not warrant us needlessly to cast ourselves down from the pinnacle of the temple, depending upon it that the angel of the Lord will bear us up, and keep us from dashing our foot against a stone, when there is another way to go down, though it be not so quick. And I would

pray that *method* in public discourses, which tends greatly to help both
the understanding and memory, may not be wholly neglected.

Another thing I would beg the dear children of God more fully
to consider of is, how far, and upon what grounds, the rules of the Holy
Scriptures will truly justify their passing censures upon other professing
Christians, as hypocrites, and ignorant of real religion. We all know
that there is a judging and censuring of some sort or other, that the
Scripture very often and very strictly forbids. I desire that those rules
of Scripture may be looked into, and thoroughly weighed; and that it
may be considered whether our taking it upon us to discern the state
of others, and to pass sentence upon them as wicked men, though professing Christians, and of a good visible conversation, be not really forbidden by Christ in the New Testament. If it be, then doubtless the
disciples of Christ ought to avoid this practice, however sufficient they
may think themselves for it, or however needful or of good tendency
they may think it. It is plain that the sort of judgment which God
claims as his prerogative, whatever that be, is forbidden. We know that
a certain judging of the hearts of the children of men, is often spoken
of as the great prerogative of God, and which belongs only to him. . . .
But to distinguish hypocrites, that have the form of godliness and the
visible conversation of godly men from true saints, or to separate the
sheep from the goats, is the proper business of the day of judgment;
yea, it is represented as the main business and end of that day. They,
therefore, do greatly err who take it upon them positively to determine
who are sincere, and who are not; to draw the dividing line between
true saints and hypocrites, and to separate between sheep and goats,
setting the one on the right hand and the other on the left; and to distinguish and gather out the tares from amongst the wheat. . . . I know
there is a great aptness in men who suppose they have had some experience of the power of religion, to think themselves sufficient to discern and determine the state of others by a little conversation with
them; and experience has taught me that this is an error. I once did not
imagine that the heart of man had been so unsearchable as it is. I am
less charitable, and less uncharitable than once I was. I find more things
in wicked men that may counterfeit, and make a fair show of piety; and
more ways that the remaining corruption of the godly may make them
appear like carnal men, formalists, and dead hypocrites, than once I

knew of. The longer I live, the less I wonder that God challenges it as his prerogative to try the hearts of the children of men, and directs that this business should be let alone till harvest. . . .

source 15

A Dislike of Doctrines
Not Savouring of
Experimental Piety

Samuel Davies

The Awakening offered Samuel Davies and others a splendid opportunity to spread New Side Presbyterianism to backcountry Virginia. However, Anglican clergymen there not only objected to the brand of religion these "hot gospellers" peddled, but they also worried about the effect of this religion upon the established Anglican Church. Samuel Davies answered some of the Anglican charges in a long letter to the Bishop of London, in whose See the royal colonies by custom fell. The controversy in Virginia went beyond the issues of revivalism and the means of grace, for Virginians came face to face with the problems of toleration and separation of church and state, problems which were not completely settled until Jefferson's Statute of Religious Liberty was enacted in 1786.

About the year 1743, upon the petition of the Presbyterians in the frontier counties of this colony, the Rev. Mr. Robinson, who now rests from his labours, and is happily advanced beyond the injudicious applauses and censures of mortals, was sent by order of Presbytery to officiate for some time among them. A little before this about four or five persons, heads of families, in Hanover, had dissented from the established church, not from any scruples about her ceremonial peculiarities, the usual cause of non-conformity, much less about her excellent

Samuel Davies to the Bishop of London, January 10, 1752, William Henry Foote, *Sketches of Virginia, Historical and Biographical* (Philadelphia: William S. Martien. 1850), pp. 190–96, 199.

Articles of Faith, but from a dislike of the doctrines generally delivered from the pulpit, as not savouring of experimental piety, nor suitably intermingled with the glorious peculiarities of the religion of Jesus. It does not concern me at present, my lord, to inquire or determine whether they had sufficient reason for their dislike. They concluded them sufficient; and they had a legal as well as natural right to follow their own judgment. These families were wont to meet in a private house on Sundays to hear some good books read, particularly Luther's; whose writings I can assure your lordship were the principal cause of their leaving the Church; which I hope is a presumption in their favour. After some time sundry others came to their society, and upon hearing these books, grew indifferent about going to church, and chose rather to frequent these societies for reading. At length the number became too great for a private house to contain them, and they agreed to build a meeting-house, which they accordingly did.

Thus far, my lord, they had proceeded before they had heard a dissenting minister at all. (Here again I appeal to all that know any thing of the matter to attest this account.) They had not the least thought at this time of assuming the denomination of Presbyterians, as they were wholly ignorant of that Church: but when they were called upon by the court to assign the reasons of their absenting themselves from church, and asked what denomination they professed themselves of, they declared themselves Lutherans, not in the usual sense of that denomination in Europe, but merely to intimate that they were of Luther's sentiments, particularly in the article of Justification.

Hence, my lord, it appears that neither I nor my brethren were the first instruments of their separation from the Church of England: and so far we are vindicated from the charge of 'setting up itinerant preachers, to gather congregations where there was none before.' So far I am vindicated from the charge of 'coming three hundred miles from home to disturb the consciences of others—not to serve a people who had scruples, but to a country—where there were not above four or five dissenters' at the time of my coming here.

Hence also, my lord, results an inquiry, which I humbly submit to your lordship, whether the laws of England enjoin an immutability in sentiments on the members of the established church? And whether, if those that were formerly conformists, follow their own judgments, and dissent, they are cut off from the privileges granted by law to those that are dissenters by birth and education? If not, had not these people a legal right to separate from the established church, and to invite any legally qualified minister they thought fit to preach among them?— And this leads me back to my narrative again.

While Mr. Robinson was preaching in the frontier counties, about an hundred miles from Hanover, the people here having received some information of his character and doctrines, sent him an invitation by one or two of their number to come and preach among them; which he complied with and preached four days successively to a mixed multitude; many being prompted to attend from curiosity. The acquaintance I had with him, and the universal testimony of multitudes that heard him, assure me, that he insisted entirely on the great catholic doctrines of the gospel, (as might be presumed from his first text, Luke xiii. 3,) and did not give the least hint of his sentiments concerning the disputed peculiarities of the Church of England, or use any sordid disguised artifices to gain converts to a party. 'Tis true many after this joined with those that had formerly dissented; but their sole reason at first was, the prospect of being entertained with more profitable doctrines among the dissenters than they were wont to hear in the parish churches, and not because Mr. Robinson had poisoned them with bigoted prejudices against the established church. And permit me, my lord, to declare, with the utmost religious solemnity, that I have been (as I hope your lordship will be in the regions of immortal bliss and perfect uniformity in religion) the joyful witness of the happy effect of these four sermons. Sundry thoughtless impenitents, and sundry abandoned profligates have ever since given good evidence of a thorough conversion, not from party to party, but from sin to holiness, by an universal devotedness to God, and the conscientious practice of all the social and personal virtues. And when I see this the glorious concomitant or consequent of their separation, I hope your lordship will indulge me to rejoice in such proselytes, as I am sure our divine Master and all his celestial ministers do; though without this, they are but wretched captures, rather to be lamented over, than boasted of. When Mr. Robinson left them, which he did after four days, they continued to meet together on Sundays to pray and hear a sermon out of some valuable book read by one of their number; as they had no prospect of obtaining a minister immediately of the same character and principles with Mr. Robinson. They were now increased to a tolerable congregation, and made unwearied application to the Presbytery of New Castle in Pennsylvania for a minister to be sent among them, at least to pay them a transient visit, and preach a few sermons, and baptize their children, till they should have opportunity to have one settled among them. The Presbytery complied with their petitions, as far as the small number of its members, and the circumstances of their own congregations, and of the vacancies under their Presbyterial care, would permit; and sent ministers among them at four different times in about four years, who stayed with them two or three Sabbaths at each time. They came at the repeated and most importunate petitions of the dissenters here, and did not obtrude their labours upon them uninvited. Sundry upon hearing them, who had not heard Mr. Robinson, joined with the dissenters; so that in the year 1747, when I was first ordered by the Presbytery to take a journey to Hanover, in compliance with the petition of the dissenters here, I found them sufficiently numerous to form one very large congregation, or two small ones; and they had built five meeting-houses, three in Hanover, one in Henrico, and one in Louisa county; which were few enough considering their distance. Upon my preaching among them, they used the most irresistible importunities with me to settle among them as their minister, and presented a call to me before the Presbytery, signed by about an hundred and fifty heads of families; which in April, 1748, I accepted, and was settled among them the May following. And though it would have been my choice to confine myself wholly to one meeting-house, especially as I was then in a very languishing state of health; yet considering that hardly the one half of the people could possibly convene at one place, and that they had no other minister of their own denomination within less than two hundred miles, I was prevailed upon to take the pastoral care of them all, and to divide my labours at the sundry meeting-houses.

•   •   •

It is true, my lord, there have been some additions made to the dissenters here since my settlement, and some of them by occasion of my preaching. They had but five meeting-houses then, in three different counties, and now they have seven in five counties, and stand in need of one or two more. But here I must again submit it to your lordship, whether the laws of England forbid men to change their opinions, and

act according to them when changed? And whether the Act of Toleration was intended to tolerate such only as were dissenters by birth and
education? Whether professed dissenters are prohibited to have meeting-
houses licensed convenient to them, where there are conformists adja-
cent, whose curiosity may at first prompt them to hear, and whose
judgments may afterwards direct them to join with the dissenters? Or
whether, to avoid the danger of gaining proselytes, the dissenters, in
such circumstances, must be wholly deprived of the ministration of the
gospel?

For my farther vindication, my lord, I beg leave to declare, and I
defy the world to confute me, that in all the sermons I have preached
in Virginia, I have not wasted one minute in exclaiming or reasoning
against the peculiarities of the established church; nor so much as as-
signed the reasons of my own non-conformity. I have not exhausted
my zeal in railing against the established clergy, in exposing their im-
perfections, some of which lie naked to my view, or in depreciating
their characters. No, my lord, I have matters of infinitely greater importance to exert my zeal and spend my time and strength upon;—To
preach repentance towards God, and faith towards our Lord Jesus
Christ—To alarm secure impenitents; to reform the profligate; to un-
deceive the hypocrite; to raise up the hands that hang down, and to
strengthen the feeble knees;—These are the doctrines I preach, these
are the ends I pursue; and these my artifices to gain proselytes: and if
ever I divert from these to ceremonial trifles, let my tongue cleave to the
roof of my mouth. Now, my lord, if people adhere to me on such accounts as these, I cannot discourage them without wickedly betraying
the interests of religion, and renouncing my character as a minister of
the gospel. . . .

. . . And here, my lord, that I may unbosom myself with all the
candid simplicity of a gospel minister, I must frankly own, that ab-
stracting the consideration of the disputed peculiarities of the established church, which have little or no influence in the present case, I
am verily persuaded (heaven knows with what sorrowful reluctance I
admit the evidence of it) those of the Church of England in Virginia
do not generally enjoy as suitable means for their conversion and edifi-
cation as they might among the dissenters. This is not because they are
of that communion; for I know the gospel and all its ordinances may be
administered in a very profitable manner in a consistency with the con-

stitution of that church; and perhaps her ceremonies would be so far from obstructing the efficacy of the means of grace, that they would rather promote it, to them that have no scruples about their lawfulness and expediency; though it would be otherwise with a doubtful conscience: but because the doctrines generally delivered from the pulpit, and the manner of delivery, are such as have not so probable a tendency to do good, as those among the dissenters. I am sensible, my lord, 'how hard it is,' as your lordship observes, 'not to suspect and charge corruption of principles on those, who differ in principles from us.' But still I cannot help thinking that they who generally entertain their hearers with languid harangues on morality or insipid speculations, omitting or but slightly touching upon the glorious doctrines of the gospel, which will be everlastingly found the most effectual means to reform a degenerate world; such as the corruption of human nature in its present lapsed state; the nature and necessity of regeneration, and of divine influences to effect it; the nature of saving faith, evangelical repentance, &c. I cannot, I say, help thinking that they who omit, pervert or but slightly hint at these and the like doctrines, are not likely to do much service to the souls of men: and as far as I can learn by personal observation or the credible information of others, this is too generally the case in Virginia. And on this account especially, I cannot dissuade persons from joining with the dissenters, who are desirous to do so; and I use no other methods to engage them but the inculcating of these and like doctrines.

•   •   •

Your lordship huddles me promiscuously with the methodists, as though I were of their party. I am not ashamed to own that I look upon Mr. Whitefield as a zealous and successful minister of Christ; and as such to countenance him. I love him, and I love your lordship, (the profession, I hope, will not be offensive) because I hope you are both good men: and if my affection to him proves me one of his party, I hope your lordship will conclude me one of your own too: yet I am far from approving sundry steps in Mr. Whitefield's first public conduct; and I am glad to find by some of his late writings that he does not approve of them himself. The eruptions of his first zeal were, in many instances, irregular; his regulating his conduct so much by impulses, &c., was enthusiastic, and his freedoms in publishing his experience to the world, in his journals, were, in my opinion, very imprudent. As to the rest of the methodists, I know but little of them; and, therefore, must suspend my judgment concerning them.

source 16

Is It a Crime for a Believer to Speak of His Having Communications Directly from the Spirit of God?

George Whitefield

During the Awakening Whitefield severely criticized Harvard College for the "deadness" in religion there. The President and faculty fought back with the charge of enthusiasm. In answer Whitefield admitted an intimacy with God which Old Light Congregationalists never could accept, confirming for them the truth of the original charge.

Boston, January 23d. 1744,5.

REVEREND AND HONOURED GENTLEMEN,
    . . . You Gentlemen . . . have thought proper to publish a Testimony against me and my Conduct, wherein you have undertaken to prove, Page 4. that I am "an Enthusiast, a censorious, uncharitable Person, and a Deluder of the People. . . ."

    "By an *Enthusiast* (you say p. 4.) we mean one that *acts*, either *according to* Dreams, or some sudden Impulses and Impression upon

George Whitefield, "A Letter to the Rev. the President, And Professors, Tutors . . . of Harvard-College . . . in answer to A Testimony Publish'd by them against the Reverend Mr. George Whitefield, And his Conduct." (Boston: S. Kneeland and J. Green, 1745), pp. 3–4, 7, 8–9, 10–11, 22.

his Mind, which he fondly imagines to be from the Spirit of God, per-
swading and inclining him thereby to such and such Actions, tho' he
hath *no Proof* that such Perswasions or Impressions are from the holy
Spirit:" This Definition of an Enthusiast (whether exactly right or not)
you are pleas'd to apply to me, and accordingly at the Bottom of the
aforementioned Page assert that I am "a Man that *conducts* himself according to his *Dreams,* or some ridiculous and unaccountable *Impulses*
and *Impressions* on his Mind" and that this is Mr. *Whitefield's* Manner is evident (you say) both by his Life, his Journals and his Sermons. . . ."

·   ·   ·

You proceed . . . to lay something more to my Charge: "Sometimes you say, He speaks as if he had Communications directly from
the Spirit of God." And is it a Crime for a Believer, much more a Minister of Jesus, to speak of his having Communications directly from
the Spirit of God? I thought that was no new Thing to the Ministers
and People in *New-England,* especially since such a remarkable Revival of Religion has been vouchsafed unto them. How are Believers
sealed; or how is the divine Life begun and carried on, if there be no
such Thing as having divine Communications directly from the Spirit
of God?

·   ·   ·

Page 8. You go on thus.—"To mention but one Instance more,
tho' we are not of such Letter-learned as deny, that there is such a
Union of Believers to Christ, whereby *they are one in him, as the
Father and he are one,* as the Evangelist speaks, or rather the Spirit
of God by him; yet so Letter-learned we are, as to say, that the Passage
in Mr. *W———d's* Sermon of the *indwelling of the Spirit,* p. 311.
Vol. of Sermons, contains the true Spirit of Enthusiasm, where he says,
*to talk of any having the Spirit of God without feeling of it, is really
to deny the Thing.* Upon which we say, That the Believer may have
a Satisfaction, that he hath the Assistance of the Spirit of God with
him in so continual and regular Manner, that he may be said to dwell
in him, and yet have no feeling of it." But *Gentlemen,* Is not this in

effect to deny the indwelling of the Spirit? For how is it possible that the Believer can have a *Satisfaction,* that he hath the Assistance of the Spirit of God with him in so continued and regular a Manner, that he may be said to dwell in him, and yet the Believer have no *feeling of it?* For my Part I cannot comprehend it. I could as soon believe the Doctrine of *Transubstantiation,* and therefore cannot retract what you are pleas'd to say contains the true Spirit of Enthusiasm, *viz. To talk of any having the Spirit of God without feeling it, is really to deny the Thing.*

.   .   .

You close your Proofs of my being an Enthusiast with these Words, "The whole tends to perswade the World (and it has done so with Respect to many) that Mr. *W.* hath as familiar a Converse and Communion with God as any of the Prophets and Apostles, and such as we all acknowledge to have been under the Inspiration of the Holy Ghost." What Tendency my Writings may have as to make People think so highly of me, I cannot determine. But this I affirm, that I would not have undertaken to preach the Gospel for ten thousand Worlds, had I not been fully perswaded that I had a *Degree* of that Spirit, and was admitted to a Degree of that holy & familiar Converse and Communion with God, which the Prophets and Apostles were favoured with, in common with all Believers. And if this had not been the Case, should I not *Gentlemen,* have lied to God as well as unto Man, when I declared at my Ordination, that I was inwardly moved by the Holy Ghost, who I believe according to Christ's Promise, will be with every faithful Minister (and so to be *felt* too) even to the End of the World.

Your affectionate humble Servant,
George Whitefield . . .

source 17

My Soul Is Grieved for
Such Enthusiastical Fooleries

Gilbert Tennent

> Not long after his return from New England, Tennent had sec-
> ond thoughts about the havoc he had caused in his own church
> and nearly everywhere he went by his rash attacks on unconverted
> ministers and encouragement of separation from them. His own
> Synod was a shambles, and he had helped to turn a number of
> mild doubters into enemies of the Awakening. His sudden turn-
> about was a surprise to both camps.

DEAR SIR,

I have had many afflicting Thoughts about the Debates that have
subsisted for some time in our Synod: I would to God, the Breach were
healed, if it was the Will of the Almighty. As for my *own* Part, wherein
I have mismanag'd in doing what I did; I do look upon it to be my
Duty, and should be willing to acknowledge it in the openest Manner.
I cannot justify the *excessive heat of Temper* which has sometimes ap-
pear'd in my Conduct—I have been of late (since I returned from
New-England) visited with much spiritual Desertions, Temptations,
and Distresses of various kinds, coming in a thick, and almost continual
Succession; which have given me a greater Discovery of myself then
I think I ever had before: These Things, with the Trials I have had
of the *Moravians*, have given me a clear View of the Danger of every
Thing which tends to ENTHUSIASM and DIVISION in the visible
Church. I think that while the Enthusiastical *Moravians* and *Long-
Beards,* or *Pickists,* are uniting their Bodies, (no doubt, to encrease
their Strength and render themselves more considerable) it is a Shame
that the Ministers (who are in the main of sound Principles of Religion), should be divided and quarrelling. Alas for it! My Soul is sick

"Extract of a Letter from the Rev. Mr. G. Tennent, to the Rev. Mr.
Dickinson of the Jerseys . . . ," New Brunswick, February 12, 1742,
The Boston Evening-Post (July 26, 1742).

of these Things: I wish that some Scriptural, healing Methods could be fallen upon, to put an End to these Confusions. Sometimes since I felt a Disposition to fall upon my Knees, if I had Opportunity, to entreat them to be at Peace. I add no more at present, but humble and hearty Salutations; and remain with all due Honour and Respect,

> Your poor worthless Brother,
> in the Gospel-Ministry,
> G. Tennent.

New-Brunswick
Feb. 12. 1741,2.

P. S. I break open the Letter my Self, to add my Thoughts about some extraordinary Things in Mr *Davenport's* Conduct. As to his making his Judgment about the *internal* State of Persons, or their *Experience* a Term of Church-Fellowship, I believe it is *inscriptural*, and of awful Tendency to rend and tear the Church: It is bottom'd upon a false Base, *viz.* That a certain and infallible Knowledge of the good Estate of Men, from their Experience, is attainable in this Life: The Practice is *Schismatical*, in as much as it sets up a *new Term* of Communion which CHRIST has not fix'd.

The late Method of setting up *separate Meetings*, upon the *suppos'd unregeneracy* of Pastors of Places, is *enthusiastical, proud,* and *schismatical.* All that Fear GOD, ought to oppose it as a most dangerous Engine to bring the Churches into the most damnable Errors and Confusions: The Practice is built upon a twofold false Hypothesis, *viz.* Infallibility of Knowledge; and that unconverted Ministers will be used as Instruments of no good to the Church.

The Practice of *openly exposing Ministers* who are supposed to be unconverted in publick Discourse, by particular Application of such Times and Places, serves only to provoke them, (instead of doing them any Good) and to declare our own Arrogance. It is an unprecedented, divisial, and pernicious Practice: It is a Lording it over our Brethren, a Degree superior to what any Prelate has pretended since the coming of CHRIST (so far as I know) the *Pope* only excepted; tho' I really don't remember to have read, that the *Pope* went on at this Rate.

The sending of our *unlearned Men,* to *teach others,* upon the supposition of their Piety, in ordinary Cases, seems to bring the Ministry into Contempt; to cherish *Enthusiasm,* and bring all into Confusion: Whatever fair Face it may have, it is a most perverse Practice.

The Practice of *singing in the Streets* is a Piece of *weakness,* and *enthusiastical Ostentation.*

I wish you Success, *dear Sir,* in your Journey: My soul is grieved for such *enthusiastical Fooleries:* They portend much Mischief to the poor Church of GOD, if they be not seasonably check'd: May your Labours be blest for that End!

I must also declare my Abhorrence of all Pretense to *immediate Inspiration,* or following *immediate Impulses,* as an enthusiastical perillous *Ignis fatuus.*

G. T.

source **18**

If This Be Enthusiasm,
Make the Most of It

Jonathan Edwards

> Jonathan Edwards believed that his friend David Brainerd, who died in 1747, was the epitome of virtue and grace. Brainerd had been very zealous during the Awakening and had made a number of enemies, particularly at Yale College, from which he was expelled for New Light activity. In order to demonstrate the compatibility between true virtue and what critics of the Awakening called enthusiasm, Edwards edited and published Brainerd's Memoirs in 1749, to which he added his own "Reflections," a part of which appears below.

• • •

II. The foregoing account of Mr. Brainerd's life may afford matter of conviction, that there is indeed such a thing as true experimental

Jonathan Edwards, "Reflections and Observations on the Preceding Memoirs of Mr. Brainerd," in *An Account of the Life of . . . David Brainerd, The Works of President Edwards* (New York: Leavitt and Co., 1851), I, 662–63.

religion, arising from immediate divine influences, supernaturally enlightening and convincing the mind, and powerfully impressing, quickening, sanctifying and governing the heart; which religion is indeed an
amiable thing, of happy tendency, and of no hurtful consequence to
human society; notwithstanding there having been so many pretences
and appearances of what is called experimental vital religion, that have
proved to be nothing but vain, pernicious enthusiasm.

If any insist, that Mr. Brainerd's religion was enthusiasm, and nothing but a strange heat, and blind fervor of mind, arising from the strong
fancies and dreams of a notional, whimsical brain; I would ask, if it be
so, that such things as these are the fruits of enthusiasm, viz., a great
degree of honesty and simplicity, sincere and earnest desires and endeavors to know and do whatever is right, and to avoid every thing
that is wrong; a high degree of love to God, delight in the perfections
of his nature, placing the happiness of life in him; not only in contem-
plating him, but in being active in pleasing and serving him; a firm and
undoubting belief in the Messiah, as the Saviour of the world, the great
Prophet of God, and King of God's church; together with great love
to him, delight and complacence in the way of salvation by him, and
longing for the enlargement of his kingdom; earnest desires that God
may be glorified, and the Messiah's kingdom advanced, whatever instruments are made use of; uncommon resignation to the will of God, and
that under vast trials; great and universal benevolence to mankind,
reaching all sorts of persons without distinction, manifested in sweetness
of speech and behavior, kind treatment, mercy, liberality, and earnest
seeking the good of the souls and bodies of men; attended with extraordinary humility, meekness, forgiveness of injuries, and love to enemies;
and a great abhorrence of a contrary spirit and practice; not only as ap-
pearing in others, but whereinsoever it had appeared in himself; causing
the most bitter repentance, and brokenness of heart on account of any
past instances of such a conduct: a modest, discreet and decent deport-
ment, among superiors, inferiors and equals; a most diligent improvement of time, and earnest care to lose no part of it; great watchfulness
against all sorts of sin, of heart, speech and action: and this example
and these endeavors attended with most happy fruits, and blessed effects
on others, in humanizing, civilizing, and wonderfully reforming and
transforming some of the most brutish savages; idle, immoral, drunkards,
murderers, gross idolaters, and wizards; bringing them to permanent

sobriety, diligence, devotion, honesty, conscientiousness, and charity: and the foregoing amiable virtues and successful labors all ending at last in a marvellous peace, unmovable stability, calmness and resignation, in the sensible approaches of death; with longing for the heavenly state; not only for the honors and circumstantial advantages of it, but above all, for the moral perfections, and holy and blessed employments of it: and these things in a person indisputably of a good understanding and judgment: I say, if all these things are the fruits of enthusiasm, why should not enthusiasm be thought a desirable and excellent thing? For what can true religion, what can the best philosophy do more? If vapors and whimsey will bring men to the most thorough virtue, to the most benign and fruitful morality; and will maintain it through a course of life, attended with many trials, without affectation or self-exaltation, and with an earnest, constant bearing testimony against the wildness, the extravagances, the bitter zeal, assuming behavior, and separating spirit of enthusiasts; and will do all this more effectually, than any thing else has ever done in any plain known instance that can be produced; if it be so, I say, what cause then has the world to prize and pray for this blessed whimsicalness, and these benign sort of vapors!

Suggested Readings

In the last fifteen or twenty years historians have shown a great deal of interest in religion in early America, particularly during the Great Awakening. Most agree that the Revival had a strong effect upon theology and the course of church history. There is less agreement about its influence in other corners of colonial society and specifically its relationship, if any, to the American Revolution. It is easy to assume that an upheaval in eighteenth-century life as profound as the Awakening had reverberations apart from religion and the churches. Concrete evidence of the effect of the Awakening upon politics, social structure, attitudes toward established churches, colonial government, and the British Empire is not as easy to find, although several recent claims have been convincing.

Following is a list of some of the significant books and articles which are attempts to make sense out of the Great Awakening or at least have shed light on it.

First, two collections of sources: Clarence H. Faust and Thomas H. Johnson, eds., *Jonathan Edwards, Representative Selections* (New York: American Book Company, 1935), brings together several of this major figure's writings; second, a recent anthology, Alan Heimert and Perry Miller, eds., *The Great Awakening: Documents Illustrating the Crisis and Its Consequences* (Indianapolis and New York: The Bobbs-Merrill Co., Inc., 1967). Because of a variety of sources, depth of treatment, and a splendid Introduction by Mr. Heimert, who quotes from the late Perry Miller's writings, it is extremely useful.

R. A. Knox has written a comprehensive book on religious enthusiasm: *Enthusiasm: A Chapter in the History of Religion* (New York and Oxford: Oxford University Press, 1950); its emphasis on the seventeenth and eighteenth centuries makes it excellent background for an appreciation of Whitefield and the Awakening. For another view of the theme stressed in the book in hand, see Maurice W. Armstrong, "Religious Enthusiasm and Separatism in Colonial New England," *Harvard Theological Review*, XXXVIII (1945), 111–40.

The only writer to give the Awakening a full treatment is Joseph Tracy, who achieved this over 100 years ago in *The Great Awakening: A History of the Revival of Religion in the Time of Edwards and Whitefield* (Boston: Charles Tappan, 1841). Good in itself, the work is enhanced by large chunks of original material which Tracy included in his pages. Other works stress either specific areas or themes. For geographical coverage, see Wesley M. Gewehr, *The Great Awakening in Virginia, 1740–1790* (Durham, N.C.: Duke University Press, 1930); Charles H. Maxson, *The Great Awakening in the Middle Colonies* (Chicago: The University of Chicago Press, 1920); and the best of these titles, Edwin S. Gaustad, *The Great Awakening in New England* (New York: Harper & Row, Publishers, 1957), which gives an excellent explanation of the theological and institutional effects of the Revival. A recent and valuable contribution to the literature is Alan Heimert, *Religion and the American Mind: From the Awakening to the Revolution* (Cambridge, Mass.: Harvard University Press, 1966). This is a large and ambitious work in which Heimert pursues the novel thesis that Calvinist advocates of the Awakening definitely influenced intellectually the Revolutionary movement, whereas their opponents, although nominally patriots, were suspicious of change and slow to act.

Perry Miller led the way in explaining the thought of Jonathan Edwards to this generation. First on anyone's list of writings about Edwards and the Awakening is his intellectual biography, *Jonathan Edwards* (n. p.: William Sloane Associates, 1949). Three of the "pieces" in Miller's *Errand Into the Wilderness* (Cambridge, Mass.: The Belknap Press of Harvard University Press, 1956) are vital to an understanding of Edwards, particularly "Jonathan Edwards and the Great Awakening" (pp. 153–66). A more conventional but good biography is Ola Elizabeth Winslow, *Jonathan Edwards, 1703–1758* (New York: The Macmillan Company, 1940).

Biographies of major figures of the period are helpful: Luke Tyerman, *The Life of the Rev. George Whitefield* (New York: Anson D. Randolph & Co., 1877), 2 vols., has many letters and documents; see also a more recent work, Stuart C. Henry, *George Whitefield: Wayfaring Witness* (New York: Abingdon Press, 1957). Isaac Backus' life was a fine example of the shift from Separate to Baptist after the Awakening. For his contribution to religious liberty and the separation of church and state, see William G. McLoughlin, *Isaac Backus and the American Pietistic Tradition* (Boston:

Little, Brown & Co., 1967). Two opponents of the Awakening were Presidents of Yale College; neither had any patience with what even suggested enthusiasm. For Thomas Clap's efforts to suppress the Revival in New Haven, see Louis L. Tucker, *Puritan Protagonist: President Thomas Clap of Yale College* (Chapel Hill, N.C.: University of North Carolina Press, 1962). Ezra Stiles was a more moderate critic of New Light ideas, and Edmund S. Morgan, in *The Gentle Puritan: A Life of Ezra Stiles, 1727–1795* (New Haven and London: Yale University Press, 1962), has written a first-rate biography of a man whose curiosity penetrated most aspects of eighteenth-century life. There is no major biography of Gilbert Tennent.

Other related works are: William Warren Sweet, *Religion in Colonial America* (New York: Charles Scribner's Sons, 1942); Leonard J. Trinterud, *The Forming of an American Tradition: A Re-examination of Colonial Presbyterianism* (Philadelphia: Westminster Press, 1949); Joseph Haroutunian, *Piety versus Moralism: The Passing of the New England Theology* (New York: Holt, Rinehart & Winston, Inc., 1932); C. C. Goen, *Revivalism and Separatism in New England, 1740–1800: Strict Congregationalists and Separate Baptists in the Great Awakening* (New Haven: Yale University Press, 1962); Perry Miller, "From the Covenant to the Revival," in J. W. Smith and A. L. Jamison, eds., *Religion in American Life*, I, *The Shaping of American Religion* (Princeton, N.J.: Princeton University Press, 1961), pp. 322–68; Leonard W. Labaree, "The Conservative Attitude Toward the Great Awakening," *William and Mary Quarterly*, 3rd series, I (1944), 331–52; Robert Sklar, "The Great Awakening and Colonial Politics: Connecticut's Revolution in the Minds of Men," Connecticut Historical Society *Bulletin*, XXVIII (1963), 81–95; and David S. Lovejoy, "Samuel Hopkins: Religion, Slavery, and the Revolution," *The New England Quarterly*, XL (1967), 227–43.

TWO HOUR BOOK

D

M

(

N

NK